PRACTICAL
DISASTER
MANAGEMENT

Colonel P. P. Marathe (Retd.)

Diamond Publications,
Pune - 411 030

PRACTICAL DISASTER MANAGEMENT

By Colonel P. P. Marathe (Retd)

First Edition : October, 2006

ISBN : 81-89724-51-5

© Diamond Publications

Cover Page :
Mr. Shyam Bhalekar

Published by :
Dattatray G. Pashte
Diamond Publications,
1691, Sadashiv Peth,
Shankarprasad Building,
Tilak Road, Pune-30.
☎ (020) 24452387

Sole Distributor :
Diamond Book Depot
661 Narayan Peth,
Appa Balwant Chowk, Pune-30.
☎ (020) 24480677

ACKNOWLEDGEMENTS

After my first book was released, I was never sure whether people would ever accept it. Clouds of doubt that existed in my mind were cleared by Mr. Dattatray Pashte who assured me that the book would be well received and that I should write another book on the same subject, purely for the benefit of students and teachers of schools and colleges. I am grateful to him for the reassurance.

For any one to write two books on the same subject is quite boring and I was dragging my feet over it. But the reason came by way of an exhibition that 'Disaster Management and Research Foundation' had organized in Pune on 26th January 2006. Though the crowd was very miniscule, each visitor spent a long time listening intently to the volunteers who had poured their hearts into the subject of Disaster Management. The visitors had gone away fully satisfied, expressing their frank opinions that the exhibition was indeed informative and most of them felt that such exhibitions should be held more frequently and reaching the masses more vigorously. These suggestions made me think again about writing this book in the simplest form, keeping it less technical. I thank these visitors sincerely. After all, this book is meant for them and their desires have to be honored.

Mr. Gopalrao Patwardhan, a "Bhishmacharya" in the field of journalism, came in my contact during the publication of my previous book on the subject. We happened to meet a few times thereafter. I realised that his love for the society's well being is so deep seated that even at this age he is full of vigour when it comes to the uplift of the masses. He has been very articulate in expressing that it was upto the people of our country to get what they richly deserved - a safe and secure life. Little did he realize that through his casual expressions of ideas he was sowing the seeds of a movement in the society. Through this book I am only joining this movement of spreading awareness among the young generation to make this nation a safer and more secure place to live in. I am grateful to Gopalrao. May God give him a long life to continue to inspire young people like me !

INDEX

PREFACE

India boasts of forward movement in technology, economy, education and social uplift. One aspect that has been neglected over the decades is that of managing actions against disasters. The awareness, technical know how and an urge to make our environment safer has to be initiated in the early stage of a human's life. The government's efforts towards including the subject of Disaster Management, as part of the curricula in schools and colleges, is laudable. This book is a small effort towards achieving this National aim.

Disaster Management is fast emerging as a mega disciplinary subject. Its applications are huge and it also has the capacity to build up value systems in the society. It encompasses philosophy and faith and structured study of causes and effects of many aspects of social, economic and environmental issues. Any effort towards studying this subject would certainly enlighten a student to start thinking holistically. Through this book, I am attempting to generate a social movement and for that to happen, it is important that the teachers study this subject and generate a deep interest in the minds of their students. They would be thus doing a great duty towards Nation Building.

Colonel P. P. Marathe (Retd)

<u>DEDICATION</u>

This book is dedicated to my deep sense of love for the humanity that pervaded my life all the time and stayed with me wherever I traveresed, be it through shrubs and meadows, the crowded streets and deserted stretches of landmass; shivering on the hill slopes and clinging to me under scorching sun or heavy downpour as well as through clear starry nights. It gave me the required warmth, radiating from within me a true, selfless and unflinching desire to rise up to every possible occasion to give a helping hand to those who needed it. It also motivated me and re-kindled my resolve to write this book.

CHAPTER 1

DISASTERS GENERAL - CAUSES, EFFECTS AND CHARACTERISTICS

General View

1.1 In this world, disasters have been occurring since time immemorial and will continue to occur. Though human is desperately trying to grapple with the disasters. he has not been able to make any significant progress in that field as yet. In fact, the stage has become more complex due to uncontrolled development everywhere. Humans are the main culprits for upsetting the ecological and environmental balance in this world. Obviously, the imbalances result in crossing thresholds of normalcy and end up into disasters.

1.2 Disaster : A Disaster is defined as *an occurrence or event that causes sudden great loss.* It is not that every disaster occurs suddenly. Today, technology has provided many tools to us to predict disasters. However, it is still not possible to accurately predict most of the natural disasters and almost all the man-made disasters. Thus, the occurrence of a disaster is quite sudden most of the time. Even when warning of a disaster is available, the loss is quite sudden because predicting the exact point of occurrence and also the accurate time, intensity and its encompassment is still quite difficult. Another reason is that our society is yet not fully aware of the connection between development and disasters. When development takes place without due considerations to environmental and geological factors, effects of disasters enhance greatly.

1.3. Connection Between Disasters & Development. Disasters cause great losses. However, disasters also prove to be beneficial in the longer run. The benefits of disasters accrue because of human's ability to learn from the experiences and put them to a better use for the benefits of those who survive. But, for those who suffer wrath of any disaster, ill effects are multifarious.

1.3.1 Ill Effects of a Disaster.

1.3.1.1 Loss of Life.

1.3.1.2 Destruction of natural and manmade structures.

1.3.1.3 Destruction of material, economic products and infrastructure.

1.3.1.4 Adverse effects on business and processes.

1.3.1.5 Health problems leading to deaths or disabilities.

1.3.2 Good effects that arise from disasters. Humans have learnt to rise from the disasters and grow better. Like, trimming of a rose plant makes the plant grow better, the humans have learnt to rise up from the throngs of disasters and grow better. The good effects are -

1.3.2.1 Land becomes fertile.

1.3.2.2 Stronger structures get built.

1.3.2.3 New systems get introduced which help more efficient management of subsequent disasters.

1.3.2.4 More vigorous development takes place in the area that is destroyed.

1.3.2.5 Brining people together, forgetting their past differences.

The following diagram sums up the connection.

Diagram showing connection between disasters & development

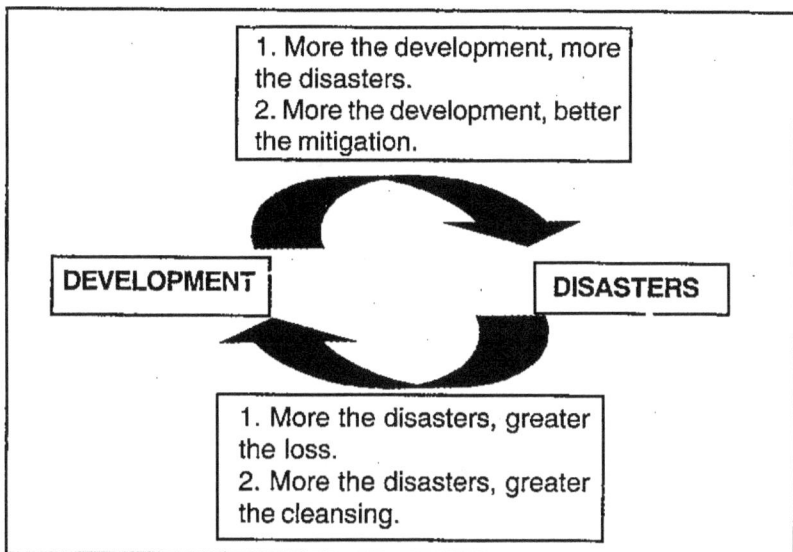

1. More the development, more the disasters.
2. More the development, better the mitigation.

DEVELOPMENT ⟷ DISASTERS

1. More the disasters, greater the loss.
2. More the disasters, greater the cleansing.

Types of Disaster

1.4 Disasters are categorized into "natural and man-made" disasters. The types in each category are :-

Natural Disasters.

1.5 These occur due to natural geological and environmental phenomena. However, the disasters do become complex due to human interventions. These disasters are :

Earthquakes, tsunamis, Floods, Cyclones, Draughts, Landslides, Forest Fires, Excessive Precipitation and Epidemics.

1.6 Earthquakes. Earthquakes are caused due to sudden release of great amount of energy that is trapped below the earth's crust. There are 14 continental plates that were formed during formation of the continents. These plates are in a continuous motion, ranging from 2 to 10 cm per year. They either move towards, away or astride one another. Whatever be the nature of the movement, it results into disturbances on the surface of the earth causing damages. The energy trapped under the crust gets released at a speed ranging from 800 to 900 km per second. The point of release of this energy below the crust is called the **Focus** and the point vertically above the focus on the surface of the earth is called the **Epicenter.** The strength of the quake that is experienced on the surface depends upon the depth of the Focus and the structure of the crust above the focus. See the diagram below :

Diagram Showing Earthquake

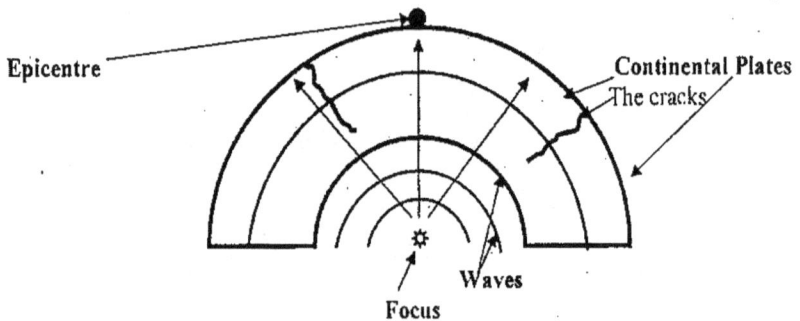

Apart from the continental plates, there are hundreds of faults and thousands of fractures in the earth's crust. The faults and fractures are the weak areas in the crust that are susceptible to the release of the energy trapped under the crust. (e.g. There is a major fault at St Andreas, United States). The effects of earthquakes are in terms of changes in the surface structure of the earth, destruction and damages to the man-made structures leading to loss of life, landslides, changing of river courses, creation of high grounds and depressions or cracks and at times discharge of molten matter of the 'Mantle' below the crust. There are secondary effects also. Earthquakes are characterized by sudden discharge of energy that traverses at great speeds. Prior to a quake, tremours are felt. Change in chemical composition of subsurface water and changes in water levels are also caused. However, technology has not yet made it possible for the humans to accurately predict the time, intensity and point of release of the energy.

1.7 Tsunami. When earthquakes occur below a large water body like an ocean, the quake waves that hit the water surface cause waves in the water body. These waves traverse great distances at an average speed on 800-900 km per hour. Because of elastic nature of water, these waves slow down. These waves have disturbances of two types - a horizontal disturbance and a vertical disturbance. When the wave is created at greater depths, the vertical disturbance, called the 'S' wave is low. This waveform is flatter when the depth of the ocean is great. The horizontal disturbance is called 'P' waves. These have a greater speed and larger wavelength. When the waves traverse closer to the shore, the depth at the continental slopes being lesser, a reactive force makes the 'S' waves taller. The height of these waves is known to be as much as 30 metres or more. See the diagram below :

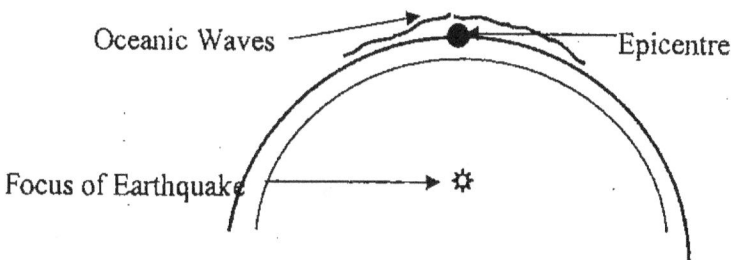

The tsunami waves strike a coast at great speed and the waves being tall, engulf almost upto 1 km of coastal belt. This depends on the ground at the coasts. There may be one or more waves. The force of the waves is so great that it destroys structures within its wake. The wave recedes with a great force and washes off the destroyed and weakened structures. This receding wave is called the 'Suction Wave'. Characteristics of tsunami that hit eight countries in Asia on 26th December 2004 was that the people observed sudden recession of seawater just before the giant wave arrived. This could be explained as a process of water being sucked while the 'S' Wave was getting formed on the continental slope. Today, warning systems are available and it is possible to warn people at least a few hours or minutes before the wave reaches a coast. The tsunami waves have been seen to be spreading inside coastlines upto a distance of 1 to 2 km. Thus, during warning time, people who can go beyond this line are sure to be saved. The receding wave is responsible to wash off the sand around and over the foundations thereby weakening the foundations and leading to collaps of structures. Thus, while constructing of houses along coastlines one needs to keep the aspect of deeper foundation in view.

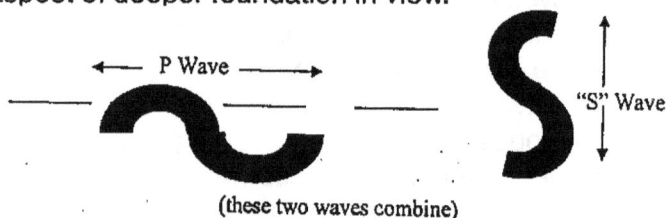

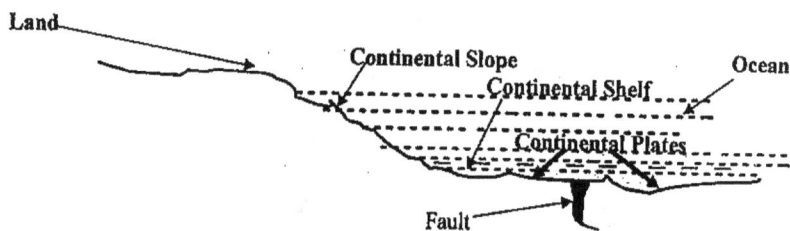

P Wave

"S" Wave

(these two waves combine)

Land

Continental Slope

Continental Shelf

Ocean

Continental Plates

Fault

1.8 Floods. Floods are regular phenomena in many countries. In Indo-Gangetic plains, the floods occur every year. In Bangla Desh, 83% of the riverine terrain gets flooded each year. Each country has a natural lay of the ground and the natural slope helps draining of water of rivers and streams. These water channels create natural river basins. Rivers and streams bring down a lot of silt with them as they flow down streams. The soils quality decides the percolation of this water in the ground. Ground within each river basin has a fixed absorption rate. Similarly, natural slopes cause specific rates of discharge of water. When rivers confluence, the lower riparian area is more at risk because of volume of water that is discharged and absorbed. The place where a river confluences a sea or an ocean (an estuary) is subjected to reverse flow during high tides, as the sea water enters the mouth of the river and does not allow free discharge of river water through the estuary. During monsoons, when the rate of accumulation of water surpasses the rate of discharge and absorption, excess water starts overflowing banks and starts spreading in low laying areas around. We call that a flood. Irrigation channels and dams help in flood control to some extent. However, many times, the rate of preciptation (Rainfall) is so great that the rate of accumulation causes pressure on the dams and water has to be discharged at greater rates. This causes floods. In cities where rainfall is heavy, natural ground slopes are few and artificial drainage systems are ill-conceived, local flooding takes place. Adhoc constructions and development affect the natural drainage of a city and blocking of gutters disallows the rainwater to get discharged. Mumbai suffers from this syndrome regularly. In July 2005, the unprecedented heavy rains (944 mm in 24 hours) caused havoc in Mumbai.

1.8.1 Characteristics. Floods do not occur suddenly. Precipitation being seasonal, prediction of floods is possible and with well installed monitoring systems and control mechanisms, it is possible to take precautionary measures. However, on the lower reaches of hilly regions, where rivers flow onto the plains, flash floods occur. These occur without warning and are quite violent. They recede as fast as they occur and the reaction time is generally restricted. In India, Himachal Pradesh, Uttarnchal and some portions of UP and Bihar face this syndrome alomost every year. After effects of any flood last for a longer time.

1.8.2 Effects. Flood waters cause destruction of weak structures, wash off property and crops, drown living beings and cut off communications marooning people for a few days. Secondary effects are felt in the form of epidemics and also electrocution through short circuiting or breaking of high-tension wires.

1.9 Volcanic Eruptions. When pressure develops over the molten magma under the earth's crust, emissions occur to release the pressure through the cracks in the earth's crust. Lava flows out along with sulphur dioxide, nitrogen and water vapour. The lava and ash solidify to make layers, over a million of years and forming volcanic mountains. Every volcano has a chamber of molten lava underneath it. Thus, the active volcanoes keep erupting time and again. Recently, a dormant volcano in Andamans has been seen to become active. Volcano eruptions affect the areas around them, burning the area around and covering habitation with ash, which is quite harmful. Warnings are available through initial small eruptions and the administration is expected to evacuate people around the volcano in quick time.

1.10 Cyclones. Cyclones are a regular occurrence on the East coast of India, Eastern and Wstern coasts of North America and the Far East Asia. Cyclones are named differently in different continents. In America, they are called 'Hurricanes', In China Sea they are known as 'Typhoons' while in North West Australia they are known as 'Willy-willies'. Cyclones also threaten Gujarat, Andhra and Tamil Nadu coasts of India. Cyclones occur so frequently in different parts of the world that they are identified by specific names. 'Katrina' and 'Rita' had taken a great toll in New Orleans, Texas and Florida in 2005. Tamil Nadu in India was also affected by cyclones in December 2005. The only cause of a cyclone is the climatic disturbance created due to temperature and pressure variations.

1.10.1 How is a cyclone created. There are various theories regarding causes and behaviour of cyclones. When air remains stationery over a land or water mass for a longer duration, that portion of air acquires almost identical temperature and humidity conditions throughout its layers. When such uniform air gets created, it is called an 'Air Mass.' The air masses may be warm or cold and may be dry or humid, depending upon temperature and moisture content. Such air masses move across huge water bodies and continents. When two air masses of different characteristics meet at any place, they do not mix easily. They have a boundary zone called 'Front.' At the fronts, these air masses exert different pressures. The temperature, moisture and pressure differences lead to creation of low or high-pressure areas and around these the air masses revolve. The low or high pressure central region of this air is called 'Eye' of a cyclone and the air either rotates clockwise or anti-clock wise. In Northen hemisphere, a low-pressured 'Eye' creates a cyclone that turns the air in Anti-clock wise direction whereas in Southern hemisphere, the air rotates clockwise. When the 'Eye' has a high pressure, it is called an 'Anti-Cyclone' and it rotates clockwise in the Northern hemisphere and Anti-clock wise in the Southern Hemisphere. The entire cyclone or anti-cyclone moves at high speeds and the lateral shift also combines with vertical movement of air current as the two different air masses try to exert pressures mutually.

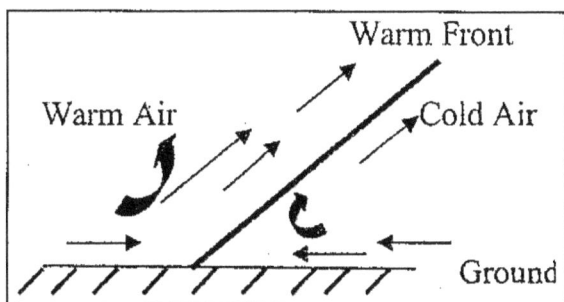

Diagram showing effects of air masses

A cyclone is said to be formed when minimum wind velocity is of 119 km per hour. Below this scale, the high wind condition is only called a 'storm' or a 'Tomado'. As a cyclone hits a coastal belt, it has two major effects - the wind pressure itself is destructive and coupled with it the oceanic waves rise up and invade landmass causing a flood condition. Thus, a cyclone is also a double trouble like the earthquakes and floods. However, in case of other disasters, one condition becomes effective first and then the second condition sets in. However, in case of a cyclone, effects of wind and waves are felt simultaneously at the coastal areas. A cyclone need not necessarily form over water bodies. Cyclones could be formed at the junctions of temperate and cold regions too. However, those formed over water bodies have a greater moisture differntial and the strength of such cyclone has been known to be greater. A cyclone is

Diagram showing a cyclonic movement

Low Pressure Eye

Diagram showing anti-cyclonic movement

High Pressure Eve

graded based on its strength - wind velocity being the major factor to decide. A grade 5 cyclone is supposed to be the strongest one anticipated with wind velocities higher than 156 miles per hour.

1.10.2 Gradations. Weather scientists Saffir and Simpson have devised a scale on which the cyclones are graded for their strengths.

SAFFIR - SIMPSON HURRICANE SCALE

CATEGORY	WIND SPEEDS			LOWEST AIR PRESSURE
	Milles / hr K	Km / hr :	Knots / hr	
1	74-95 :	119-153 :	64-82	980+
2	96-110 :	154-177 :	83-95	989-965
3	111-130 :	178-209 :	96-113	964-945
4	131-155 :	210-249 :	114-135	944-920
5	156+ :	250+ :	136+	Below 920

1.10.3 In New Orleans, the city is below the Mean Sea Level of that place. To protect the city, an embankment, called 'levee,' was created around the city. The embankment had been constructed to withstand a cyclone of about grade 4. However, hurricane Katrina was of grade 5 and the pressure of the seawater breached the embankment, causing flooding of New Orleans and resulting into heavy casualties.

1.10.4. Today, it is possible to predict a cyclone and its strength and direction, thanks to the weather satellites. Thus, a greater reaction time is available to the administration to place mitigation measures. Even then, it is not possible to eliminate damage to structures in a short time frame. At best, about three to four days of notice is available to the administration. In case of 'Katrina', inspite of a warning casualties did occur.

1.11 Draughts. Draughts is a common and regular feature all over the world. Change in ecological conditions is causing indifferent rains at many places. Desertification is taking place due to incessant cutting of trees. Though governments are desperately trying to launch irrigation projects and efforts for water harvesting, the results are far from encouraging. This needs to be taken up as a social drive. During draughts, enough warning is available and any government can take effective measures with the help of NGOs and local help groups. Draughts adversely affect crop production due to scarcity of water. Water and food shortage result into deaths due to hunger and thirst. To prevent draughts and manage them, a deliberate development effort is required by the governments and people. Water harvesting and irrigation canal development as well as practical solutions of linking of rivers could be resorted to.

1.12 Landslides.

1.12.1 Landslides are mostly caused as secondary effects due to earthquakes, tsunami, cyclones, floods or heavy precipitation (rains). They are also caused due to natural denudation and soil erosion, depletion of tree cover due to uncontrolled deforestation and natural cracks that get developed due to weathering effects. Development of land on the hill slopes during construction of new roads or building complexes weaken the hill structures. Such areas are prone to landslides. Such areas become vulnerable with other natural calamities like earthquakes, floods and heavy rains and chances of landslides increase manifold. In 2005 August, after Konkan belt of Maharashtra experienced multiple landlides, a survey was done and it was noticed that many hills had developed verticle cracks in Sahyadri region, on the western slopes. Thus, that region would be threatened with landslides.

1.12.2 Landslides are a local phenomenon within a limited area. The degree of landslide depends upon the soil condition, the slope and rock structures and expanse of the hills. Red soil is easier to slide when not firmed in. Soil helps to hold rocks and the rocks help to harden the soil. Tree roots help in retaining the soil and thus, landslides get restricted due to the trees. Due to heavy rains, soil gets washed off and the rocks become loose and slide down the slopes. These rocks also bring down soil around them and trees get uprooted. The falling debris causes damage to the structures located on the slopes and many a time life and property gets buried. After the rains of July 2005, an area in Mumbai suffered landslide destroying about 140 huts that were constructed as temporary structures and could not resist the falling mass of rocks. The soil of the hill that suffered from landslide in Mumbai was red and not firm. Tree cover in the area was limited. Almost all the huts had no foundation and were constructed without authorization and certainly not as per building norms. Thus, it is seen that even when landslides are a natural phenomenon, the destruction is caused due to unregulated human actions. In Raigad district, a complete hill (Jui Dongar) collapsed during unprecedented rains of 26th July 2005, burying hundreds of people underneath. Landslides cut off road communications and people trapped in it suffocate and get injured due to falling debris. Rescue is difficult as landslides leave a loose mass of earth. The more one digs, more the quantum of loose soil from heights starts collapsing. If landslide results after heavy rains, the rescue becomes even more difficult as the mud continuously drains down the slopes.

1.13 High Precipitation. High precipitation in some areas has been experienced due to environmental conditions. Some people attribute it to the effects of 'El -Nino', a disturbance over the Pacific Ocean that has been affecting regions as far as the Indian Ocean. This is causing cascading effects on the weather pattern. The global warming has also affected the weather patterns. Whatever be the exact reason, heavy precipitations, particularly the rains, cause hovoc in some parts. The weather satellites do indicate such conditions well in time. However, exact quantum of precipitation can never be predicted. Such a precipitation causes landslides, local flooding and marooning of people. There are effects like building collapses and short-circuiting also.

1.14 Epidemics. In developing and underdeveloped countries, the standard of living of many people is so low that their environment and general living conditions are unhygienic and result into epidemics. Malaria, Influenza, Cholera and hepatitis are some of the most common epidemics experienced in these countries. The World Health Organization (WHO) and various social groups do work in conjunction with the concerned governments. However, the requirements far outmatch the available resources to face these diseases. Lack of education and perpetual poverty are the main basic reasons. In India, about 30% population lives in substandard conditions and non-availability of safe drinking water and lack of medical facilities add to the death-rate.

Man-made Disasters

1.15 These occur due to causes that are purely attributable to human actions. The frequency, types and intensity of such disasters is increasing due to the development and socio-political systems that the world has evolved. These disasters are :

1.16 Accidents in Transportation Systems. In India, every year at least 85,000 to 1 Lac people loose their lives because of road accidents and damages caused have been estimated to be in the tune of Rupees 55,000 Crore. (Reader's Digest January 2006). The rail, air and marine accidents also add to this figure of losses. Most accidents related to transportation systems are due to human errors and some due to system failures. During travel on two wheelers, many riders think that wearing helmets is not convenient. It is not the convenience that matters. Even cricketers wear helmets to keep away from injuries. The momentum of impact created from collision of two vehicles is 1000 times more than the momentum at the impacting of a cricket ball. Bad road surfaces and wet surfaces increase the degree of probability of an accident. But most accidents occur due to human errors and neglegence.

1.17. Fires. The major losses are due to fires that are caused due to human errors and system failures. Fire accidents are very common and are avoidable. Even if they are not totally avoided, the losses could be brought down appreciably if the humans are aware and are prepared. Fires can be explained with the help of a diagram as shown below. For a fire to start and sustain, three elements are highly essential - Combustible material, oxygen and heat. To extinguish any fire at least one element out of the three should be cut off from the point of fire. The entire fire- fighting theory is based on this fact.

Fires are categorized as class 'A' fires that are caused when a material like paper, cotton, grass, wood etc burn; Class 'B' when the combustible material in question is oil based like Petroleum products, rubber, paints and any other oil; Class 'C' fires are due to other chemicals including cooking gas and Class 'D' fires are when metals burn - like sodium, sulphur etc.

Depending upon the type of fire, the effects differ and the methods of mitigation vary. There are materials that are non-combustible - like sand, flour, synthetic foam etc. These can be used to suffocte combustion. There are some materials that are fire resistant - these materials have a very high Ignition Temperatures or Fire Point. Such materials are used for protection purpose. To avoid fires, one should take care to ensure that gas knobs of the gas cylinders are switched off when not in use, ensure no sparking takes place in electric connections and electricity loads are within the limits. Do not burst fire crackers on roads or near car parking and also closer to a building. Also, ensure that oil lamps are not left burning, unattended.

1.18 Explosions. With the onset of terrorism, explosives have attracted great attention. Terrroist organizations use explosives to create Improvised Explosive Devices (I.E.D.) to cause damage and destruction and rein havoc amongst the population. The reason for such a commonality is the easy accessibility of such people to materials that can be used for causing explosions. An explosive is a substance that burns vigorously when ignited, causing such rapid chemical changes that a great amount of mechanical force is created causing a bursting effect. This bursting or explosion creates three types of effects - blast effect, shattering effect and heat effect in the surrounding area. Explosives are generally categorized as - High explosives, Low explosives and miscellaneous composition like smoke and incendiary. The category depends upon the velocity of detonation. This velocity varies from 2 km to 9 km per second. The effects are caused mainly due to the immense velocity. The energy that gets released after the explosion traverses in a wave-form in three dimensions. Some of the high explosives are - TNT (Tri-Nitro Toluene), RDX (Research and Development eXplosive), Amatol, Baratol, Gun Powder, Gun Cotton, Nitro Glycerin and Nitro Cellulose etc. Some of the substances that are used as initiators and are also in high explosives category are Sulphuric Acid, Lead, Antimony compounds, Phosphorous compounds, and Mercury compounds. Explosives are brown, dark grey, pale yellow or off white in colour. They can be in solid, liquid, jelly and gaseous forms. Most of them being non-metallic are difficult to get detected by metal detectors.

1.18.1 During war, explosives are used in bombs and missiles, rockets, grenades and mines. These explosives have different kinds of fuses. For an explosive to function, a flash is required for ignition. A fuse creates a flash that gets passed onto a detonator that magnifies the flash and spreads it in the body of the explosives packed in a container. There are many types of mechanisms in fuses - combusting material mechanism, pressure and pull mechanism, release and percussion mechanism and electric or electronically operated mechanism. There are mechanisms that get activated by tremours. Terrorists use crude mines and bombs - these are called 'Improvsided explosive devices' or I.E.Ds. An I.E.D. is a device that has explosive packed in a case resembling an article of daily use - like brief cases, parcels, chocolate boxes etc. The main principle behind it is that an observer should make a mistake in identifying the I.E.D. and the explosive should go unnoticed. Thus, innovative packaging methods are used. Letter bombs are also used, however their effect is restricted to a very limited area and only the person opening a letter containing an I.E.D. suffers. Blasts in trains and buses in Mumbai were caused by the I.E.Ds. that went unnoticed.

1.18.2 Intensity of an explosion depends upon the quantity of explosive used and the type of explosive as well as the compactness of packing. Greater the compactness, greater is the detonating velocity and thus, greater the intensity of explosion. Effects of an explosive increase when an explosive is placed in a confined place. In open areas, the detonating wave gets dissipated easily. Explosives have a lesser effect on irregular surfaces and composite walls. Their effect also decreases if the explosive is wet. Certain material like sand helps in absorbing the blast wave. Rubber and cork are good protectors against splinters and also help in absorbing the blast wave shocks. Explosives cause casualties mainly due to blast waves. Blast waves shatter surfaces of objects. The next effect is due to splinters that embed in bodies. Effect due to heat is the least because of greater dissipation of the heat energy. Explosions due to gas cylinders also fall in the same category. Secondary fires may start due to explosion. The 9/11 attack on the World Trade Center in the US caused a double pronged damage - one, due to the impact of the planes on the structure and second, due to explosions of the fuel in the planes, causing greater damages to the already damaged and loosened structures.

1. 18.3 The general public must ensure that an unattended package / bag / parce / should not be touched and should be reported to the authorities. Where a suspect item is identified, the public should vacate the place beyond danger area, without panic.

1.19.Nuclear, Biological & Chemical Disasters (NBC)

1.19.1 As the mankind has progressed, the usage of such agents has increased tremendously. Apart from the usage during war, accidental leakages have occurred in the past. When one thinks of war, scenes of Hiroshima and Nagasaki flash in the minds. With teorrist organizations getting more modern in their methods, there could always be a chance that any of the NBC agents/means could be used for attacks. Thus, one needs to also know about these.

1.19.2 Nuclear Energy Related Issues. Nuclear energy is the most potent of the known ones. This energy is released by fission of an atom or fusion of more atoms. A combination of fission followed by fusion and the reiterative process multiplication could compound the amount of energy that can be obtained. In nuclear reactors, these processes are controlled with due care. When used for peaceful purposes, its effects are very beneficial. However, when used without safeguards and for destructive purposes, the effects could be devastating. The main effects are caused by Alpha, Beta and Gamma radiations. Human body can accept certain dosage of radiations without any ill effects. However, when the radiation level to which a human body is subjected crosses that threshold, ill effects start occurring. Beyond a particular stage, the effects are irreversible. This is the reason why radiologists are more at risk and also the cancer patients who are subjected to radiation therapy are given calculated doses under control. Following issues need consideration :-

1.19.3 Explosion of an Atomic Weapon. Atomic weapons, when exploded could have devastating effects due to great speed of atomic fission or fusion or a combination of the two with 'critical mass' of a kind that results into production and release of great amount of energy. This energy gets released suddenly in different forms - light, heat and blast waves and also in the form of radioactive rays. The strength of each of the energy type depends upon the mass of radioactive fuel subjected to fission or fusion. The first nuclear bomb dropped on Hiroshima on 6th August 1945, codenamed 'little boy' had a fuel that was of the size of a tennis ball. Bomb making is quite technical, costly and difficult. Today, due to proliferation of technology and availability of material, a probability exists when such weapon systems - albeit, in crude form, may land into the hands of fundamentalist organizations. Also, national interests of many countries may lead to an international situation when one country having such weapons may use it on the adversary. The threat of usage of neclear weapons has grown over the past few decades. Thus the need to be aware and prepared for such eventuality.

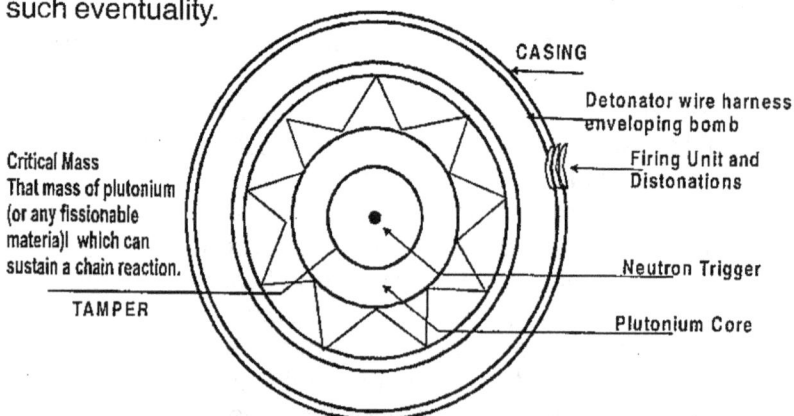

Critical Mass That mass of plutonium (or any fissionable materia)l which can sustain a chain reaction.

TAMPER

CASING

Detonator wire harness enveloping bomb

Firing Unit and Distonations

Neutron Trigger

Plutonium Core

Diagram : the inside of a bomb and its working

1.19.4 How the Mechanism Works. Explosives are detonated to produce symmetrical shock waves travelling inwards. This compresses the plutonium core to a critical mass. A trigger sends a stream of neutrons that cause a chain reaction in plutonium. The plutonium atoms split, releasing enormous energy in a short time.

1.19.5 Atomic weapons could be exploded above the earth's surface, at a great height or may be very close to the surface - a low airburst. It could be exploded on the surface or below it and even under water. The point on the surface of the earth immediately above or below the point of explosion in called 'ground zero'. Different bursts have different effects and the decision to effect a particular type of burst would depend upon the destruction intended, technology of the delivery system and ease of execution. From the viewpoint of adversity of effects on the human a low airburst on a populated area could possibly be the worst kind. An atomic explosion has the following after effects - a blinding flash of such an intensity is produced that it could cause damage to retina causing temporary or permanent blindness. Along with such a flash, heat is produced that has the capacity to vapourise most of the material within a few hundreds of meters of radius. Outside this radius, many living and non-living materials would be burnt beyond recognition and outside a radius of about 1 km, fires would ensue, causing further damage. The blast effects would be even more shattering. Within a radius (say 1 km), all structures would be erased to the ground. For the next few km, there would be 90% damage to structures and similar percentage of people dying instanteneously and the damage percentage would come down with each increase in radius. Damage template would depend upon the strength of the bomb and the point of burst and also the capacity of the human habitat to withstand it. In short, if a class 'A' city with a population of 10 million and a population density of 1000 people per square km is considered then on an average, there could be 12000 people dying instantaneously, about 100 thousand people dying within a few minutes, another 200 thousand dying due to secondary effects of building collapses and getting buried under rubble. Others would suffer from radiation effects and would die a slow death due to various radiation burns. These figures are only rough estimates. As said earlier, the effects would be largely dictated by the strength of the bomb and the height of burst. Weather conditions - including precipitation, wind direction and wind speed play thier parts in residual effects. Immediately after the burst, a mushrooming cloud occurs, consisting of dust, smoke and air particles. This cloud rises up and starts drifting with the wind. Radioactive particles drift with the cloud and settle down on the earth's surface causing radiations. Effects of immediate radiations in the area surrounding the 'ground zero' are also felt. The energy released is converted in various forms as per the following distribution - 15% Light, 20% Heat, 45% Blast and 20% radiations (5% immediate and 15% residual radiations).

1.19.5.1 Blast Wave Effects : Blast effect causes maximum immediate damage. A blast wave causes great initial pressure, moving away from the Ground zero and causes buildings to collapse or suffer serious damage. It has a shattering effect. A vacuum is created around the point of explosion and after the wave peters off, a suction wave gets generated and causes reverse pressure. The reverse pressure causes damaged buildings to shatter further and tumble. Though the forward wave is speedier, the reverse wave acts upon damaged structures with considerable force.

1.19.5.2 Radiation Effects : The radiation effects are due to Alpha, Beta particles and Gamma rays. Alpha particles are nothing but atoms of Helium and are positively charged. Beta particles are like the orbital electrons and are negatively charged. Gamma rays are not particles. These are the electromagnetic radiations like the X-rays and have great penetrative power. While the Alpha particles are not penetrative and can be stopped by even a thin paper, the Beta particles can be stopped by a thin sheet of metal. The gamma rays penetrate through 36 inches of earthwork, wood and concrete and also through all living beings causing extensive burns. Any dose that is administered on a human body beyond 60 Rads per hour causes long-term effects. A dose of 600 Rads (total) is fatal. The amount of radiation energy received by a body is measured by dosimeters.

RAD : It is the absorbed dose of 100 ergs of energy per gram of exposed body. This is called 1 RAD.

ROENTGEN : It is a unit of radiation exposure dose. It is measured in 'R' per hour.

REM : It is the Roentgen equivalent of biological damage due to radiation.

Note : The exposure is generally measured in terms of RAD

1.19.6 Accidental Radiations : Chernobyl nuclear reactor of USSR developed accidental leak of nuclear radiations and caused havoc in the country in 1986. During Tsunami of 26th December 2004, there was a possibility expressed that Kalapakkam nuclear reactor of India could suffer some cracks. However, the reactor was safe. But, with more nuclear energy for peaceful purposes being harnessed, location of such a plant has to be very carefully planned. In case of tsunamis and earthquakes damages could be caused to such plants causing leakages.There have been hundreds of accidental leakages from nuclear reactors, in the world so far. Many of them have taken place in the USA. safeguards are essential.

1.19.7 It had been reported earlier that Al Quaida had plans to produce suitcase bombs. Such a device, if detonated, could have catastrophic effects on human beings. This shows that the hazard of nuclear devastation can not be ignored by any country in the present day world scenario of energy and technology acquisitions and geopolitical environment. Thus, stricter checks and restrictions need to be imposed.

1.19.8 Biological Hazards : Like the nuclear hazards, biological hazards are also quite dreadful. These hazards are increasing day by day. It is the proliferation of technology coupled with lack of control or desire to cause damage and destruction that makes this hazard potent. India is one of the signatories to an international pact through which many countries have agreed to refrain from producing biological weapons. Biological weapons are comparatively easier to handle and deliver and are not very bulky. Their detection is indeed difficult and spread is moderately speedy and more localized than nuclear weapons systems. Following table gives some of the most potent biological agents and their effects. Scientists are researching newer agents each day.

AGENT	CHARACTERISTICS AND EFFECTS	FATAL DOSE
Anthrax	It's a bacteria, Causes fever, septic and shock, results in difficulty in breathing. Death occurs within 24 to 72 hours.	1 to 2 milligram (mg)
Botulinum Toxin	Causes blurred vision, difficulty in swallowing, paralysis, respiratory failures and death within 24 hours	Less than 1 mg
Aflatoxin	Causes hemorrhage, convulsions, coma, liver cancer and slow death.	---

1.19.9 Chemical Agents : Bhopal gas tragedy can be considered as the severest of all accidents caused due to chemicals. The cyanide content of the gas produced in the factory leaked in December 1984 causing thousands of deaths and many more to suffer permanent damages to eyesights and respiratory systems. Chemical weapons have been known to be used since the Second World War. Hitler killed thousands in gas chambers. It is claimed that Saddam Hussein, the deposed dictator of Iraq, had used chemical weapons against some Kurds and even during Iraq-Iran war. This report is certainly yet to be authenticated. Many countries including India have committed to non-use and development as well as storing of chemical weapons. However, disasters caused due to industrial leakages can never be overruled. Following table gives details of some of the most potent chemical agents.

AGENT	EFFECTS
Vx nerve gas	Disrupts nervous system, causes convulsions, respiratory failures, paralysis and death.
Sarin	Attacks nervous system and can cause death within minutes
Mustard Gas	Causes skin and eye burns, blisters and lung diseases and often cancer

Living With Disasters

1.20 Disasters result into destruction of structures, burying and injuring lives, damage to property and vegetation, deaths due to drowing, burning, suffocation, diseases and hunger. Most of it is avoidable. One disaster may cause secondary disasters. These are the cascading effects. Human errors and habits have come up as the biggest factors increasing effects of disasters. These errors and habits need to be effectively corrected. People and property has to be insulated from the effects and preventive measures need to be instituted.

1.21 Human has faced many natural disasters in the past. The degree of destruction has been dictated by the development and human interventions. Of late, human has emerged as the biggest enemy of own self and that of the environment on this globe. The disasters get compounded due to human actions. The Mumbai floods and the Leptospirosis that caused deaths were due to suppression of the nature and unhygienic conditions in which humans lived. Piles of trash and blocked drainages caused diseases. Lack of awareness and callousness in maintaining good living conditions, uncontrolled constructions and non-adherence to building norms continue to compound human woes. Mankind has to live with the disasters. Only way to survive is through Preventions, Adequate Warnings and Awareness, "Hazard-Vulnerability-Risk Assessments and effective Remedial Measures and more importantly, the 'Will' to help own self and others. We need to create a healthy environment so that the present generation can survive well and leave a healthy legacy to the subsequent ones. We also need to be conscious of our value systems to build strong and correctly designed structures, using the right material. For this to happen, Nations and Communities have to work together consciously and set right the systems, pervading beyond the barriers of classes and clans, the regions and religions, languages and genders. After all, it's a war- a War against the disasters.

EXERCISES FOR PRACTICS

Q1 : Why do the earthquakes take place ?

Q2 : What are the types of floods and how do they occur ?

Q3 : What is a Focus and what is an Epicentre ?

Q4 : What is the difference between a cyclone and an anti-cyclone ?

Q5 : Write a detailed note on radiation effects of a nuclear explosion.

Q6 : Why does development affect disasters ?

Q7 : What should the society do to prevent road accidents ?

Q8 : When an explosive is detected in a public transport, what actions would you take to ensure greater safety of all passengers ? Write in about 100 words.

Q9 : What care will you take while celebrating a festival, to ensure better safety ?

❖ ❖

CHAPTER 2

HOLISTIC VIEW OF DISASTER MANAGEMENT

Disaster Management as a Mega Discipline

2.1 : In the previous chapter, we saw that when disasters occur, they have multiple effects. Prevention and mitigation of disasters are continuous processes that are repetitive and involve many organizations, people and nations. The multifaceted nature of the disasters warrants a broader view. The following diagram represents such a view :-

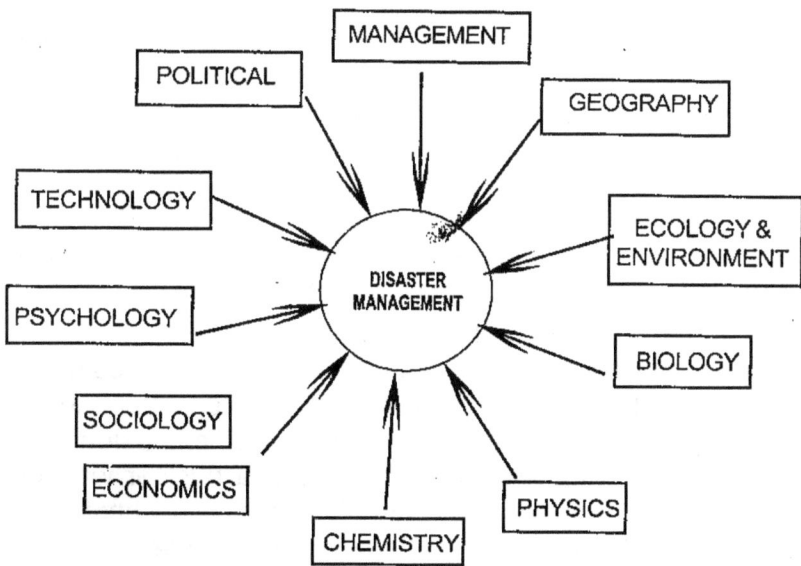

2.2 Geographical Interventiions. All the natural disasters are caused by geographical factors. Be it earthquakes or floods, Tsunami or Volcanoes, Draughts or Landslides. Geography plays a major role. Even for man-made disasters, the effects get enhanced or diminished due to geographical factors. For example, during wars, undulated areas suffer lesser damage in bombardments. Here, the geographical factor of terrain (ground) offers a greater degree of natural protection. Also, effects of nuclear holocaust are felt lesser in an area with greater vegetation.

2.3 Ecological and Environmental Factors. Human's effort to develop economically, industrially and socially has caused environmental degrades and has affected ecology. Some of the disasters are due to these factors. These factors also have to be taken into account while planning mitigation strategies and solutions.

2.4 Biological Interventions. When any disaster occurs, saving of lives by application of medical and biological interventions becomes a very major task. Depending upon the type of disaster and the characteristics of a casualty, medical and biological interventions are decided and implemented. When we consider biological weapon systems, the preventive measures are different. For chemical agents, there are different antidotes. For burn injuries, a different treatment is given and similar is the case for epidemics and water related disasters. The time of intervention and the degree to which biological interventions are available decide the final losses in life.

2.5 Physics Related Interventions. Disasters are a set of forces. The forces could be in the form of water, forces under the crust of the earth, forces of gravitation when a landslide or building collapse occurs or the forces of explosion of a conventional explosive or even burst of a nuclear weapon system. The warning systems are also based on technological advancements that are physics based. The mitigation efforts too have physics intervention in various forms. Thus, study of disaster cannot escape study of physics.

2.6 Chemical Interventions. When we talk about disasters like fire, it is the chemistry of combustion that one needs to understand. The working of explosives also falls in this category. The preventive measures are based on study of chemistry and the antidotes of chemical weapon systems are also based on the study of chemistry. Many protective clothing against Nuclear, Biological and Chemical agents and fire resistant suits are a result of study of chemistry and polymers.

2.7. Economics. Disasters cause losses not only in terms of life and material, but also harm infrastructure, lead to the losses in business and warrant gigantic efforts towards reconstruction, rehabilitation and further development. All this costs money. It has been observed that the cost of compensation and rehabilitation is almost 1.5 times the cost of assessed damage. When reconstruction is added to this cost, the total cost is almost 1.8 times that of the assessed damage. The total cost does not include additional avenues of employment that have to be created for the displaced victims. The costs would also differ depending on the area affected by disasters and the basic economic condition of the country. The economic condition also dictates the speed at which the rehabilitation and reconstruction can be done. Apart from this, the government and other mitigating agencies have to be created, administered and activated to face disasters. This subject is being introduced at different levels of academic studies and various officials also have to be trained. The cost of education is considered as a part of preparatory phase. A Nation creates and deploys warning systems against disasters. All this costs money and is a burden on the economy. Budgetary allocations are now being done to take this aspect into account.

2.8 Sociology. Disasters enshrine many social issues. For example, in India we celebrate many festivals. The society culture has transformed into a showy one where the increased pomp and show and the uncontrollable crowd offer an invitation to disasters. Take the case of stampede in Kumbha Melas or during Haj Pilgrimage; study the accidents arising from indiscriminate use of firecrackers during Diwali or the extravaganza during Ganesh or Durga festivals. They all pose an increased threat of disasters. The divisive tendencies in the society trigger terrorist activities. Poverty gives rise to unhealthy living conditions causing epidemics. Lack of social awareness creates ecological imbalances. All these are social causes of disasters. These are avoidable. Disasters also create social issues. Disasters cause deaths and many children are orphaned, many families are devastated and loose their means of livelihood. Their rehabilitation is a major responsibility of the government in particular and the society in general. Disasters also bring people together, making them forget the old grudes and work beyond the cracks of religions, castes, economic diversities and regionalism.

2.9 Psychology. Psychology plays a major role during disasters. Psychology of different cateogry of persons involved with disasters can be a separate subject of study. It can be studied under different heads as follows :

2.9.1 Psychology of a Victims. While disaster is imminent or actually occurs, most victims show a natural fear. Their actions under fear are not logical and may cause grave problems. However, it is seen that a small number of victims also show great courage and presence of mind. They come forward to save others. It has been observed that those who are trained show greater balance and more structured actions in the face of a disaster. Immediately after the main fury of disaster recedes or when secured at safe place, the survived victims show tremendous gratitude towards their Gods followed by grief. Many younger people are angry at the fate they suffer but mostly there is a wide spread of grief and resignation. This is generally followed by an eageness to receive as much aid as possible and thereafter an urge to maximize out of the grave situation grips them. Children show deep insecurity for a while and if correctly tackled, come out of it faster. Old people show a more resigned attitude. There are many shades of psychology that are displayed by the victims.

2.9.2 Psychology of NGOs & Other Agencies Attempting Mitigation. Mostly, the members of NGOs show a great sympathy towards the victims and go all out to help people. However, most of them have a short-lived enthusiasm. Though every kind of small help goes a long way towards helping the victims in restoring their lives, many who do not activiely participate for a long time and offer short one-time intervention have been observed to be doing it more for self-satisfaction than for helping the victims. Though it is wrong to generalize this statement, observed behaviour of many people falling in this category confirms this. To varying degree, an NGO or any other agency does attempt to highlight their work. Probably, it is natural.

2.9.3 Psychology of Civil Administration. Officers and staff of civil administration have been found to be of two broad categories - one, who are totally dedicated and involved with great compassion and sense of responsibility and second, the ones who just wish to play safe and not risk own interests. The second category also shows some degree of indifference, which goes up as the work pressure increases. One reason of this behaviour is lack of training and preparations and inability to work positively under pressure.

2.9.4 Psychology of Military and Para-Military Forces. While the Armed Forces show a great degree of élan, sense of responsibility and promptness and professionalism, the para military forces differ from one type of force to the other. Notwithstanding the degree of effectiveness, all of them certainly show a great positive attitude and a deep sense of duty. This appears to be possible due to build up of a mental frame through continuous training.

2.9.5 Psychology of Political Parties. It has been repeatedly observed that the members of the ruling party try to maximize their ability to help out and satisfy the victims as speedily as possible, the members of the opposition party attempt to outdo the party in power and also go out of their way to point lacunae in the actions of the party in power. Both actively attempt to woo the media. There is far lesser cooperation between the political parties. In our country we need to rectify this anomaly and work together, for the benefits of the victims.

2.10 Technology. Following points merit considerations :

2.10.1 Engineering aspects are extremely important prior to, during and after the disasters. The strengths of various structures dictate survivability and resulting damages. Designs, geology of places and building norms matter a lot. The degree of damage depends upon not only individual structures but the general development norms followed. To highlight this issue one must know that the constructions created on hill slopes have to be differently constructed than those situated on the coasts and should be different from those located in flood and earthquake prone areas.

2.10.2 Technological advancement of a country coupled with economic conditions of people dictate the survivability. Technological advancements allow a country to deploy more reliable and efficient warning systems, thus allowing the peole and the administration a greater reaction time to take loss-prevention measures of timely evacuation of people.

2.10.3 Technology is also very important in the field of reconstruction and subsequent development. The town of New Orleans had a protective 'bund' called levies to keep the encroaching seawater away. However, the Hurricane of 2005 breached the old levies. New Orleans administration should have strengthened the levies long back.

2.11 Political Issues. These issues are double-edged weapons. If correctly used, they would help in fighting a disaster unified. Most of the time, political disagreements and rivalries create hindrance in relief work. Political interference in rescue and relief operations greatly hinders the tasks. For effective mitigation, political will and unification of efforts are essential. Even at International Levels, diplomatic strains affect mitigation efforts. At the other end of the canvas, Disaster Mitigation helps cementing of differences and even creation of stronger diplomatic bonds.

2.12 Management Issues. These are at the core of the sphere of Disaster Management. It is through management aspects that all other issues could be directed, controlled and activated. We need to see the management aspects in greater details separately, in subsequent paragraphs.

Management View

2.13 Disaster Management, though a multi disciplinary and multi-faceted function, is well handled through the right management techniques especially applicable in this field. Disaster management is more akin to warfare management. It also has all those aspects of management and warrants mobilization of resources and efforts at National Level. In warfare, the complexities are many and yet, an enemy is known. Strengths, weaknesses and possible actions of an enemy can be well anticipated and appropriate answers could be found at military, diplomatic and economic ends at strategic, operational and tactical levels. In comparison, a disaster is an unseen and mostly unpredictable enemy and the entire human race is more or less reactive to any disasters. Apart from natural causes, the human actions escalate the ill effects of disasters and render a society helpless. This makes Disaster management a complex and challenging field. A 100% safeguard is only an illusion; however, minimizing of liabilities forced by disasters is a bounden duty of each individual, group, organization and the nation as a whole.

2.14 Many of the times people are known to mix up the two terms - Disaster Management and Disaster Mitigation. We need to have clarity in meaning and encompassment of both these terms.

2.15 Disaster Management. Disasters can be managed at different phases. Each phase has several sub-phases which we can call 'stages.' If we compare any disaster with an event, then we can view them correctly. Any event has a preparatory phase where we prepare to execute the event. The next phase of any event is the actual execution when the event is conducted. The third phase is when we take all actions to wind up the event by taking stock of material and finances and close the event. If a cricket or a football match is an event, then the teams are selected and they practice for the match. They also plan their strategies. All this is part of the Preparatory - Phase. On the day of the match, the players actually play and dynamically attempt to win using all their skills and tactics, their power and coordination. This is called the Execution Phase. After the match, we take a stock and see where we stand and make changes to our future team compositions and strategies. This is the Post Event Phase. Disaster management also goes through these phases. These phases are : Pre-disaster Phase (Preparatory Phase), Active Disaster Phase known as 'During Disaster Phase' (or 'Execution Phase') and Post Disaster Phase. The phases are explained here under and also indicated in the diagram. These phases partly run sequentially and partly simultaneously, when we consider a multi-disaster scenario. Every major disaster invariably triggers a secondary disaster. Thus, while one phase of the first disaster is being processed, some other phase of another disaster is processed. While we fight a disaster through Execution Phase, we simultaneously prepare for the next one, whether both these disasters are connected or not. These processes are repetitive and the end point of one process often merges and overlaps beyond the starting point of another phase. These phases and their stages are explained in the subsequent paragraphs.

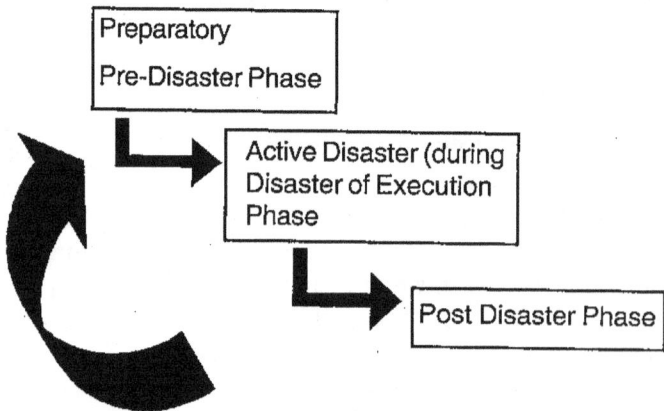

2.15.1 Pre-Disaster Phase. Pre-Disaster Phase is very important because, the degree to which planning and preparation is done in this stage helps in mitigating a disaster and bringing down the losses. It has the following stages :-

2.15.1.1 Anticipation Stage. This stage involves assessing the likelihood of natural and some manmade disasters, their likely probability of occurrence, the frequency of occurrence and likely intensity and their effects. This is called 'Threat-Vulnerability and Damage Assessment.' It is also called Risk Analysis. This helps the people and the administration to take actions of suitable preparedness and many other aspects of development planning.

2.15.1.2.1 Planning Stage. This stage is very important. Planning mainly involves the government and other agencies that are involved in disaster mitigation. There are many things that the government has to plan. The plans are regarding prevention and mitigation (reduction in liabilities by saving life and property during disasters and actions to restore normalcy) of disasters. This aspect is dealt with in greater details in the next chapter.

2.15.1.3 Preparation Stage. Preparations involve implementation of preventive measures, preparing the society and preparation of development plans. It also involves deployment of warning systems for different disasters. The preparedness leads to enhancement of warning time, greater survivability and most imortantly, to reduce reaction time. Again, this aspect will be dealt with in detail in the next chapter.

2.15.2 During Disaster or Execution Phase. This phase has two major stages as under :-

2.15.2.1 Rescue Stage. This stage starts either when a prior warning is available (like in case of floods and cyclones) or when there is no warning available before a disaster occurs (like in case of an earthquake). If warning is available, the rescue involves removing people to saftey and securing life and property before the disaster strikes. This can rarely be achieved to a 100% level. The rescue efforts after a disaster strikes involves rescuing people from danger and providing immediate medical and stabilization help to them and salvaging of as much material as possible. More details have been given in the chapter on Rescue.

2.15.2.2 Relief Stage. Relief is an activity where the victims who are evacuated prior or during a disaster are rehabilitated at a safe location and they are helped to lead a normal life and rehabilitated permanently with due compensation and aid. This stage involves offering of shelter, food, medical help, loss and damage compensation and all administrative actions that would help the victims to restart their lives from the losses suffered by them during disasters. This aspect is also dealt with in detail in the subsequent chapter on Relief.

2.15.3 Post Disaster Phase. In this phase, the stages are invariably overlapping and have to be handled with great efficiency. The stages of this phase are as follows :-

2.15.3.1 Rehabilitation. Rehabilitation can be seen in two parts - firstly, settling the displaced victims into relief camps in the form of temporary rehabilitation and secondly, constructing and repairing houses to permanently settle them, with due loss and damage compensation. The first part has to ideally commence immediately on receipt of warning and the second part should commence immediately after the disaster is controlled. The second part takes time to commence because, the government has to take many decisions concerning place of resettling the victims, the degree to which the settling is to be assisted and financial effects of the entire activity. The government also has to collect data regarding the exact damage and destruction and the amount of compensation required. The disbursal of compensation is time consuming.

2.15.3.2 Reconstruction. Reconstruction can be viewed as reconstruction of private properties in which the government assists the victims and secondly, the reconstruction of damaged infrastructure like bridges, roads, railway lines and ports etc.

2.15.3.3 Development. After any disaster, the government has to undertake development of the devastated areas and also re-plan further development in a manner that the area does not suffer a similar devastation in case a similar disaster occurs. New development norms are laid down, building laws are amended, risk assessment is re-done and accordingly new systems are developed and deployed.

2.16 Disaster Mitigation. This term relates to the activities that fall in the implementation category. While Disaster Management activities encompass the entire gamut of the three phases, the mitigation encompasses the last two phases. The better the management, greater are the chances of reduction in liabilities. However, this can only happen if the management correctly executes the mitigation and takes timely and accurate decisions, keeps a good control and directs all the efforts adequately. In mitigation, predisaster phase is not actively included. Also, the development stage of post-disaster phase is not really part of mitigation. We can thus say that mitigation forms part of overall management and is the implementation face of the disaster management. The following diagram may clarify the issue.

RELATIONSHIP BETWEEN DISASTER MANAGEMENT AND MITIGATION

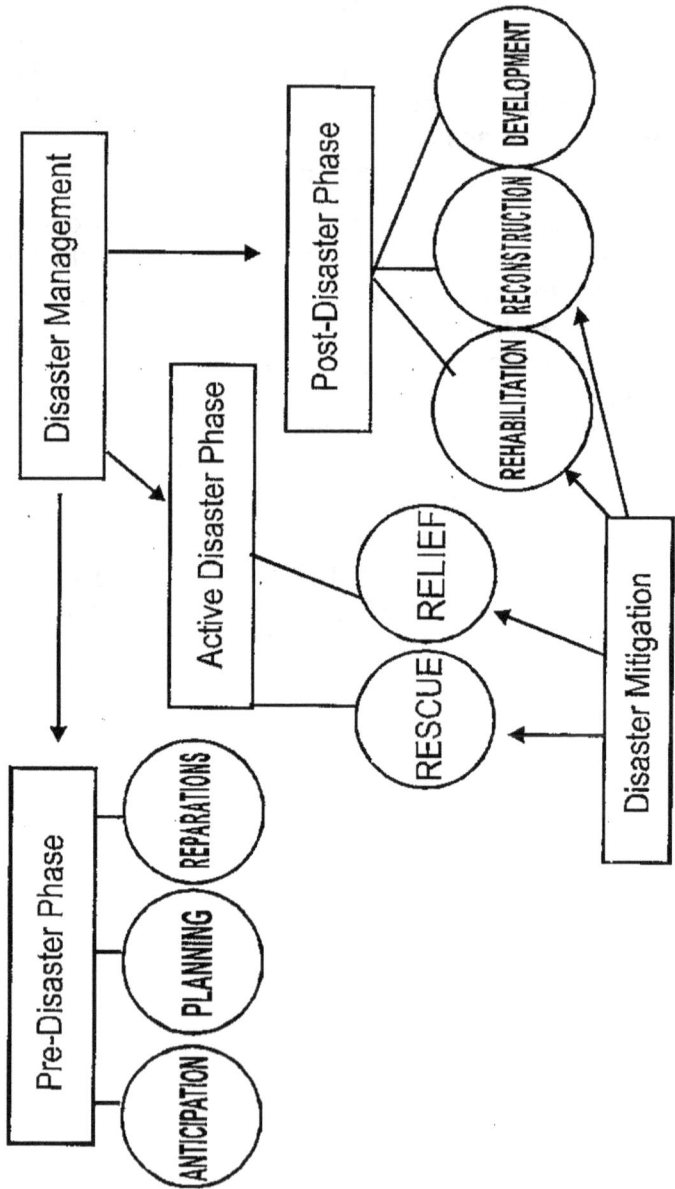

EXERCISES

Q1 : Why is Disaster Management a mega-discipline ?

Q 2 : Why is technology related to disaster management ?

Q 3 : When does execution phase start and what are its stages?

Q 4 : Why is the pre-disaster phase important ?

Q 5 : Explain the psychological aspects of a disaster in 100 words.

Q 6 : How do the political aspects affect a disaster ? What should be done to ensure smoother execution of work during disasters through political angle ?

Q 7 : Explain the role of technology in minimizing damages caused by disasters.

Q 8 : How will you go about assessing disaster related risks in your town ?

Q 9 : Write a 200 words note on the social issues that could be involved in Disaster Management, based on the idea that has been generated in the above chapter.

❖ ❖

CHAPTER 3

ANTICIPATION (ESTIMATION), PLANNING AND PREPARATIONS

3.1 : If disaster is an event, then, Disaster Management is also an event comprising several sub-events. Disasters are pro-active whereas the management and mitigation are broadly reactive. If disaster is like a blow from an opponent, then Disaster Management is a reaction that involves preventing the blow and ensuring that even if the blow strikes the injury is minimal. In a boxing bout, every boxer anticipates the opponent's blow and avoids it or blocks it on the guard. A cricketer wears a protective helmet and a pad. The hockey players wear shin guards. It is the characteristics of any disaster that a human cannot take proactive counter-blow like a boxer hitting back the opponent or a general planning a counter offensive. In disasters, the counter offensives are in the form of greater anticipation and protections and more efficient mitigations.

3.2 Disasters are categorized on four levels as follows :-

3.2.1 L0 Level Disasters. These are local accidents causing damage and destruction to life and property and need immediate mitigation. Road accidents, drowning while swimming, small fires, localized food poisoning etc fall within this category. The number of people involved is less, the area affected is small and the reprecussions of damage are restricted only to the families whose kin are affected. The reacting force need not be necessarily from the administration and local people present at the site are the main actors in mitigation. Though, in case of fires, people do take help of fire brigade. Police force is invariably involved.

3.2.2 L1 Level Disasters. These are local disasters also, however, their magnitude is much higher. More number of people are involved and efforts required to mitigate are more deliberate and long lasting. Though spontaneous reaction is required, this is invariably not available at the location of the disaster and a gradual build up takes place. Administration gets involved at this level and even specilized agencies are deployed. These incidences are like a major fire in a locality, industrial accidents, major building collapses, stampede or any other accidents during celebrations, rail and air accidents etc.

3.2.3. L2 Level Disasters. These disasters have a higher intensity and may cover an area that is wider and more number of people get affected. There is a requirement of more intense mitigation efforts by the government administration including use of specialized forces. Such disasters are like riots, grave industrial accidents, terrorist attacks etc.

3.2.4 L3 Level Disasters. These disasters are so intense that it covers several towns and villages and districts. The intensity warrants State, National and at times International level mitigation efforts. Large number of people get affected and the degree of destruction is severe and renders a large population ineffective. Earthquakes, Floods, Tsunami, Cyclones, War, Nuclear Accidents are some such types of disasters.

3.3 Anticipation (Estimation). Though the disasters are categorized as mentioned above, there is a definite need to further analyze them. The main reason for this analysis is that it provides us with information about the probability, frequency and intensity of their occurrence with respect to geographical coverage and then leads us to decision making about planning of mitigation efforts at all levels and also the degree of preparedness. The process of such analyses brings out the risks that exist and may give an idea of the likely destruction. It helps in preparing the society to face disasters and bring down liabilities. In chapter 1, we saw how cyclones have been graded based on intensity of the winds. Similarly, earthquakes are graded on Richter scale and are indicative of the strength or intensity of the quakes. It is possible to lay down such scales for each type of disaster. We need to understand a few terms that would help us estimate or anticipate disasters and their effects.

3.3.1 Threat. Many experts in the field of disaster management have given various definitions of 'Threat.' However, for a common person to understand, we need to draw a platform of a condition of disaster. When we say that a disaster is an occurrence or event that causes sudden great loss, then, a threat could be defined *as the onset of a condition that may cause a deviation to the state of normalcy of life.* The condition could be viewed in terms of probability of its occurrence, the frequency at which it could occur and the intensity of its occurrence. When such a condition crosses a threshold of acceptance, it can be termed as a hazard and its effects are seen. *Thus threat is the possibility of occurrence of disaster cuasing a degree of loss/damage.* In order to be prepared for a disaster, threat of a disaster needs to be identified and then further analyzed.

3.3.2 Vulnerability. Vulnerability Analysis is the second step towards determining likely effects of any disaster. Vulnerability can also be called as "Weakness" that results into some loss. If strength is the capacity to withstand any adversity, then vulnerability is the reciprocal of the degree of strength. It reveals the percentage of adverse effect that could be caused related to a particular intensity of disaster. The aim here is not to apply mathematical or statistical formula, but view it from a commoner's eyes. Say, if strength (or capacity) of a structure is defined arbitrarily as 100% if it could withstand a shock of an earthquake with 7 on the Richter scale and epicenter within 1 km of the structure, then, its vulnerability can be taken as 0%. But, the same structure may not be able to withstand a quake of 9 points on the Richter with epicenter within 1 km. In such a case, if the structure gets damaged to 30% extent, then its capacity to withstand is 70% and so, vulnerability is 30%. This indicates that vulnerability or capacity to withstand cannot really be measured on any absolute scale. The capacity to withstand has to be related to the intensity of the threat. Or, simply put, vulnerability is directly related to the intensity of a disaster (not mathematically directly proportional). Thus, for defining vulnerability, Intensity should be the basic parameter.

Thus, vulnerability can be defined as the degree of destruction that a structure or an area or human population is likely to suffer for a given intensity of a disaster. It is inversely proportional to the strength to withstand. Vulnerability of the same structures and areas and population may be different for different disasters.

3.3.3 Risk. While vulnerability denotes a broad degree of destruction, the risk spells out specific damage in absolute terms that may be caused in a particular area. To give example, one can derive the deductions regarding a flood situation in terms of specific villages that are likely to get affected, the buildings that may collapse, the number of people who would be rendered shelterless and the amount of damage to crops and also infrastructure. This absolute number or quantum is important for making plans and preparing for the mitigation.

3.3 Planning. To face any disaster, one needs to plan actions much prior to occurrence of the disasters. The planning involves many agencies, which need to establish congruence in disaster preparedness. Disaster Management has to be viewed as a holistic system, synergising the preparations and for that a high degree of centralized planning is essential. Planning involves many things. We shall see the most essential aspects of the planning in the subsequent paragraphs.

3.4 Creating Organizations for Mitigation. To perform Disaster Management tasks effectively, a nation has to have an organization to look into planning and managing the entire aspects. A nodal agency has to exist within the government at each level like the National Level, State Level, District level. National Disaster Management Authority has already come into force in 2005. This agency is responsible for planning, coordinating, controlling and directing all the functions. A similar sub-set of this is expected to come into force at state and district levels. Apart from this, organizations have to be created and administered to actually operate for conducting rescue and immediate relief duties. The central and state governments have to also address the post-disaster rehabilitation, reconstruction and development activities through their normal channels. All these agencies have to coordinate with one another within a given framework and accepted policies, under a centralized command. Over and above the officially created organizations, many Non-Government Organizations (NGOs) also participate in the mitigation activities. A good coordination has to be established with these also.

3.5 Policy Formulation. The central government is responsible for formulating policies concerning all three phases of disaster management. The policies include :-

3.5.1 Policy on Vulnerability and Risk Assessment. This will include conducting of the time bound surveys with given periodicity and assessing the likely damages.

3.5.2 Policy on Creating a Fund for Disaster Management. In India, a fund has now been created to address the needs of creating organizations, deployment of warning systems and compensation package as well as expenses on executing mitigation tasks.

3.5.3 Policy on Development and Construction Norms. To increase survivability, these policies are very important.

3.5.4 Policy on Creating and Deploying Warning Systems and Communication Channels. Implementation of these policies offer greater reaction time and cuts down activation time, apart from offering better flexibility and reaction of all organizations.

3.5.5 Policy on Preparedness and Awareness of General Public. These would include educational norms concerning disasters, training and operational norms of industries, institutions and controlling authorities of infrastructure (like dams and installations). These would also lay down safety and security norms for individuals to follow under different circumstances.

3.6 Information Management Plans. In order to react speedily, execute mitigation efficiently, have correct vulnerability and risk analyses and also for accurate damage estimation, a voluminous data is required to be compiled and managed. The administration has to plan and implement the information management plans dynamically. The information management involves collection of certain static data during non-disaster phase and certain dynamic data during 'Active Disaster' phase. The information and data (data is the analyzed information) include the following :

3.6.1 Geographical and Geological Data. This data is very important from the viewpoint of vulnerability and risk analyses. The data would also help in prediction of some of the disasters. It would also help in identifying safe areas for evacuating the victims. The geographic data that also includes demographic data would help to identify the population risks and behavioural pattern as well as the natural resource data would help in planning rescue methods.

3.6.2 Industrial Data. This would help in identifying the likely industrial hazards and availability of mitigation resources and also coverage of vulnerability of the populace around the industries.

3.6.3 Weather Data. Weather data is a dynamic data, which would help in prediction of weather, related disasters.

3.6.4 Infrastructural Data. This data helps in identifying threats and vulnerability of the infrastructure and also probable use of this infrastructure. This data would also help in planning mitigation methods for each area. Availability of Hospitals and their capacity, availability of routes for transportation and evacuation, availability of safe structures to house various elements of the mitigation efforts form part of this data.

3.6.5 Resource Related Data. This data would help in identifying resources like engineering, medical, food and clothing items availability, supply chains and their time of activation.

3.6.6 Apart from this, survey data of revenue nature, development data, agricultural data etc should also be compiled.

3.6.7 Data regarding co-opting agencies, their capabilities, contact information etc would provide the government with the ability to coordinate the mitigation processes with these agencies and help speedy and effective deployment.

3.7 Planning the Command and Control Systems. Execution phase can only be successful if the command and control systems are planned correctly. The commnand and control systems involve establishment of control centres at different levels, at different locations and correctly staffed and equipped. The control centres could be static at National, State and District levels but should be of mobile nature at lower levels when disasters strike. Depending upon threat and vulnerability assessment of different disasters in different areas, placing of such control centres have to be planned.

3.8 Planning the Location of Rescue and Relief Organizations. Depending upon the threat and vulnerability analyses, prior positioning of the mitigating agencies (organizations) and their responsibilities with practical alternatives has to be planned. Thus, a unit of CRPF earmarked for disaster mitigation duties and co-located with a district headquarter may have alternative tasks in other districts. Thus, various combinations of deployment of units would form part of a centralized coordinated plan.

3.9 Resource Planning. To mitigate any disaster, tremendous amount of resources have to be mustered up. The resources have to be identified, created, acquired and deployed. Any delay in deployment of resources hamper mitigation efforts. The administration has to plan holding of resources and decisions have to be made regarding their holding levels, places of holding and holding organizations and their locations. Administration has to also identity sources from where some important resources could be made available at short notice. The resources may be in the form of medical aid manpower and medicines, food, clothing and shelterning resources and distribution agencies. Rescue and engineering equipment has to be placed with the rescue units and additional resources have to be mustered up. Transport management plans have to be kept ready and coordinated. Management of all those resources cannot be done without preparing resource management plans for different resources, during different disasters at different locations.

3.10 Rescue Planning. Though it is difficult to plan rescue at higher levels of administration, this should be done by the rescue units, for different disasters, at different locations based on vulnerability and risk analyses. This would lead to the resource planning as well. It would also lead to task allotments of all rescue related organizations.

3.11 Relief Planning. Relief is an activity that gets triggered the moment a warning is issued or when any disaster actually strikes. Relief planning could include identification of location of relief camps, structures that could be used during disasters, planning to enhance the holding capacity of hospitals within the region and creation of static and mobile relief teams. Identification of NGOs and other social organizations that could coordinate with the government to execute relief tasks also forms part of this planning. A rehabilitation plan could also be made in the outline.

Preparations

3.12 Success of any operation, be it a war or any event or even disaster mitigation depends upon detailed planning and rigorous preparations as much as correct execution. It would not be wrong to state that ultimate success of disaster management depends upon the success in mitigation aspects. Thus, through the right management correct mitigation is ensured. 'Preparedness' forms an inescapable part of Disaster Management. Since the government is duty bound to ensure safety and secruity of the citizen of a country and also for that of the infrastructure, material - public or private, it forms a major player in the act of preparedness. However, the onus of being prepared for every eventuality also rests on every individual, each organization and group within their area of jurisdiction and activities. After all, to uphold the Safety, Security and Integrity of the country is every citizen's fundamental duty. The elements of safety, security and integrity imply that we the people are resonsible for own and other's safety. Any loss and damage to life and property causes great burden on the nation. Our government spends crores of rupees each year towards disaster mitigation. This is an unwarranted liability on the nation and any reduction in such liability would allow the country to divert the resources spent on mitigation efforts towards other important development avenues. One needs to view individual responsibility from this viewpoint. Thus, preparations encompass a wide range of elements - the government, the private and public sector organizations, industry, institutions, formal and informal groups and individuals. This aspect has been dealt with in the subsequent paragraphs.

3.13 Preparations by the Government. Ths government has to prepare in many aspects as follows :-

3.13.1 Creation of Responsive Command and Control Infrastructure. The ogvernment has to create administrative command and control structure and work out its triggering. Since this infrastructure may be at times common with the normal functioning of the government, the preparations must include identifying and briefing of those officials who would form part of the emergency command control structure.

3.13.2 Warning Systems. Identifying and Deployment of warning systems and connecting the systems down to the lowest level of villages is an important part of preparations. In India, we have weather satellites that can give warnings to weather stations, regarding cyclones and precipitation. The seismology centres that exist can detect quakes but it is not possible to give prior warnings of quakes. In so far as floods are concerned, the observations on rains and accumulation of water resulting in water levels do give flood warnings at Taluka and District levels. Tusnami warning systems have come into existence in the world. These systems record energy releases and wave patterns and can give warning of tusnami a few minutes to a few hours prior to the tusnami striking a cost. However, warnings regarding other disasters are not yet possible. Even when warnings are available, in India, the issuance of warnings at lowest levels is still not well structured. Ideal warning system is indicated in the diagram below :-

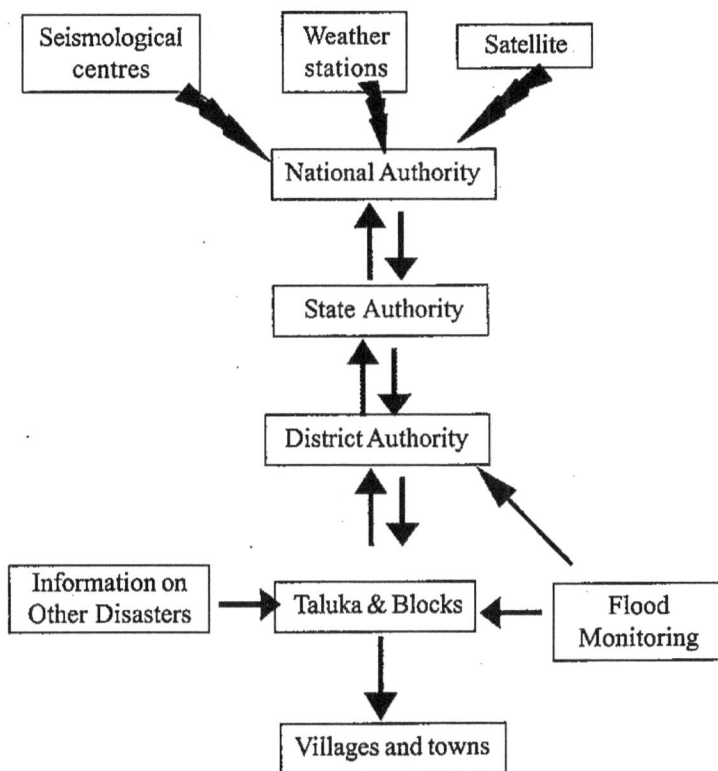

(Through Police and Village Sarpanchs)

3.13.3 Resource Holding. Resources that are required by the government for rescue and relief are of different categories. Some are expendable and have to be planned and collected at the last moment. But, all the rescue material and some of the relief material, which is of non-expendable, type needs to be held to some extent at all times. This material is required to be held and maintained by some organization. The material has to be also located with great thought on its transportability and deployment.

3.14 Preparations by Industries, Corporate Offices and Edcuational Institutions. Each of them has the following preparations to do :-

3.14.1 Construct building infrastructure to be safe during natural disasters.

3.14.2 Arrange safe and secure working of people and machinery and assess safety norms and procedures from time to time.

3.14.3 Make disaster mitigation plans of own institutions / offices to include forming of teams and arranging resources for mitigation (like fire fighting equipment), make and rehearse evacuation plans.

3.14.4 Arrange for regular training of all personnel to include fire fighting, organizing safe stay of workers / inmates, first aid and casualty evacuation and elementary rescue.

3.14.5 Conduct rehearsals of mitigation plans periodically.

3.15 Hospitals. Hospitals have a great responsibility. Not only that they have to be prepared for receiving emergency casualties and accommodate extra strength, but they also have to think of management of the patients when the hospital premises themselves face disasters. It is a critical situation when a hospital faces any disaster - be it an earthquake, a fire, a gas cylinder explosion etc. Their disaster mitigation plans have to be worked out in a manner that the patients are well catered for. Following points merit consideration. :-

3.15.1 The procedures when patients are being operated when a disaster strikes.

3.15.2 Disposal of patients in the ICUs and the CCUs.

3.15.3 Disposal of patients who cannot walk and go to an alternate safe place.

3.15.4 Control over oxygen cylinders that are in use.

3.15.5 Radiation room management.

3.15.6 Shifting all the patients to alternative locations, particularly in case where patients are critical or not attended by any relatives.

3.15.7 Arrangements to accept additional casualties and their treatment.

3.15.8 Evacuation of patients stuck in the lifts.

There are many issues that the hospitals have to consider and put in place an organization - based on shifts and train their staff in handling medical emergencies while facing emergencies of disasters.

3.16 Places of Entertainment. Cinema and concert halls are not really geared up to face disasters. A study of fire episode in Upahar Cinema Hall of Delhi is a live example. Break out of fire could be avoided if due care is taken. In India, there are no periodic rehearsals of evacuation of the public from any place of entertainment. This needs to be made obligatory. Though, cinema halls do display fire extinguishers, it is generally noticed that the staff is not trained to handle the equipment. Cinema halls and other theatres must announce the standing orders for emergencies before any show begins. Making these rules is the responsibility of civic administration. With terrorism showing its ample presence in India and elsewhere, such orders become mandatory. Availability of well-equipped first aid kits is another area that needs looking into.

3.17 Religious Places and Celebrations. The orders for religious places should be same as for cinema halls. Here, the difference is that during festivities, crowd management is by and large tardy, even chaotic. The rate of inflow is not commensurate with the rate of outflow of the devotees. Use of oil and lamps is at times so negligent that chances of an accident are high. Stampede is another syndrome that needs to be controlled. Strict adherence to safety norms should be imposed. Long queues outside religious places are also problematic. Fire that raged at Mandhardevi temple in Satara district of Maharashtra is an example of how slip-shod the administrative arrangement was made. Presence of small time vendours adds to the woes and there is no control over them at all. Sanitation facilities are never upto the mark during such big gatherings and onset of an epidemic is always on the cards. Take example of Kumbha Mela. The water in which holy dip is performed turns quite harmful by the second day of celebrations. During Kumbha Mela at Nashik in Maharashtra in 2003, lack of space created maximum problems and the police were not geared up (not withstanding the tall claims of the administration prior to the festivities) to face acute emergencies. Ganesh festival or Durga Pooja require better handling everywhere. In fact, the concepts of erecting temporary structures that occupy roads leads to uncalled for traffic congestions and create greater hazards. This practice needs to be immediately reviewed. The temporary structures could be erected on open grounds. There is no aversion to religious celebrations; however, the functions should ensure public safety and avoidance of inconvenience to the public. There are no fire precautions taken by most of the organizations celebrating such festivities. Also, we need to change our mind set as per the changing lifestyles and the demography. These should be true for any celebration - be it a political rally or a social function. During festivities, roads are used for bursting firecrackers. This creates tremendous hazards to the congested traffic and needs to be totally stopped. Every year, firecracker shops and factories report fire accidents. This could be very much avoided through strict regulations. For all this to happen, the administration and the judiciary need to study the matter and pass bills and laws in the larger interest of the society. After all, every unwanted accident imposes load on the aministration and causes damages.

EXERCISE

Q1 : What are the different levels of disasters and what are their administrative effects ?

Q2 : Explain vulnerability and risk analysis.

❖ ❖

CHAPTER 4

ORGANIZATIONS INVOLVED IN DISASTER MANAGEMENT

4.1.1 : Before we proceed towards study of further aspects of disaster management, we must see some of the organizations that are involved in it. Most of the organizations are of the government while there are a few Non-Governmental Organizations (NGOs) involved in this field.

The National Level Organization

4.2 At National Level, the **Home Ministry** has been tasked to handle the issues concerning Disaster Management and nominated as the Nodal Agency. This ministry is responsible for coordinating all aspects of disaster management on behalf of the government. It is responsible for creating structures by taking cabinet approvals and floating any bills regarding the Disaster Management. It is also responsible for creating, mobilizing and directing all required resources, obtain and give sanctions for various activities and coordinate with the other government ministries and the state governments. It is also responsible to accept and offer disaster related aids and assistance to and from other international agencies and states. To assist the government in deciding policies, advise on implementation and all disaster related matters, a National Disaster Management Authority (NDMA) has been formed. In November 2005, a National Disaster Management Bill has been passed, obtaining legislative sanctions to authorize the NDMA to formulate policies and identifying basic structures and operational laws in connection with Disaster Management.

4.3 National Disaster Management Authority (NDMA)

This body is the strategic advisory body that keeps control over Disaster Management aspects in the entire country. It not only acts in an advisory capacity, but also provides policy back-up, observations, monitoring and operational interventions at the Centre and State levels. This Authority is chaired by the Prime Minister and has members who have vast experience in their respective fields of specially. The NDMA has formulated and identified structures and forces at the National Level as well as state levels for dealing with disasters and a fund has been set-aside for this purpose. Presently, a force level of eight battalions of CRPF (Central Reserve Police Force) has been earmarked especially to deal with the rescue and relief work during national disasters.

4.4 State Disaster Management Authority (SDMA). This is a sub-set of the NDMA, at the state level. The process of establishing such state level authorities is in progress. The state level authorities are likely to be chaired by the respective Chief Ministers. The Vice Chairman of a State Authority is likely to be the Chief Secretary or any such suitable person so appointed. The members and advisors are likely to be appointed by the Chairman. At the state level, this authority would perform the same functions as the one performed by the NDMA at the National Level.

4.5 District Disaster Management Authority (DDMA). Chaired by a District Collector, this authority is the main operating government agency to manage disasters. It is at this level that rescue and relief operations are expected to be planned, executed, monitored and controlled. The main feed back of situation and progress is collated and forwarded by this agency. It is through this agency that the national level operations are made effective.

4.6 All the above agencies have to establish control centres for efficient planning, directing, controlling and executing disaster management activities. Apart from these controlling organizations, the government deploys many other organizations during disaster management, for actual operational part of rescue and relief, followed by rehabilitation and reconstruction. Let us see these organizations.

4.7 Government Organization Involved in Rescue and Relief. The following organizations are deployed by the government on the basis of need and availability :-

4.7.1 Police. The police organization is used for the tasks concerning radio communications for warning and disaster related command and control purpose. Actually, there is a need for having a separate and dedicated organization for communications. Police is also deployed for security of the area affected by disasters, to restrict access and to prevent looting. Police also gets involved in identification of the dead and handing / taking over of dead bodies after due documentation and subsequent disposal. This force, if correctly trained and equiped, can be tasked as the first respondents.

4.7.2 Home Guards. At the district level, the district collector controls the home guards. They are used for further beefing up the policing force for cordon and security duties and some of the trained manpower is used in rescue and immediate relief activities as well as debris and road clearance. There is a need to train and equip this force to perform rescue operations.

4.7.3 CRPF and CISF. The Home Ministry controls these forces. These have been very effective in rescue, excavation and providing immediate relief. These forces still do not have any specialized equipment as of now. However, through their training and discipline, they have performed disaster related tasks with great élan.

4.7.4 BSF. The Home Ministry also controls this organization. However, this organization is generally used during disasters in border areas. The BSF is used for heavy rescue duties and light relief duties.

4.7.5 Armed Forces. The armed forces are used for heavy rescue duties including specialized difficult missions, evacuation of casualties by air, naval divers for search etc. The air force also launches supply and airdrop missions. Army is the main arm for heavy and specialized rescue and higher level of relief and initial rehabilitation. The Defence Ministry controls them and the missions are launched on requisitioning by the state governments. The army has a permanent system where the operational and peace units are allotted Internal Security (IS) duties where the entire country is divided into various areas and sub-areas and units are permanently allotted their tasks and are made to establish liaison with the civil administration periodically to update disaster threat analyses, resource planning and operational plans. Thus, they are highly organized to render help during disasters. The resource availability, quality of training and well established systems and speed of response make their performance effective.

4.8 Non-Government Organizations (NGOs) Involved in Rescue. There are a very few NGOs that are involved in rescue duties and are registered for this purpose. However, these organizations have to act to supplement the other government organizations. These organizations being not so well equipped as the armed forces, they have to undertake simple rescue or act as second string to the government organizations. Some of the Organizations are Disaster Management and Research Foundation (DiMaRF) in Nashik and Pune in Maharashtra, Bharatiya Jain Sanghatana (BJS) and many other smaller organizations. These organizations are very useful in clearing landslids, debris clearance and cllection and disposal of the dead. These organizations are very handy in providing relief material to people who are stranded. They have the capacity to establish relief camps that are run with the government support and they also help the victims in psychological recovery. Some religious organizations also come forward during such times. There are a number of organizations which, though not registered for this purpose, manage to collect and distribute relief material and run camps.

4.9 International Organizations. There are many International Organizations that come forward during severe disasters. Some of them are - the Red Cross, OXFAM etc. These organizations have been rendering great service in terms of provision of medical aid; relief material and many organizations are engaged in special rescue missions. The efforts of these organizations are coordinated by the Central

Government.

Rehabilitation and Reconstruction

4.10 The major responsibility of rehabilitation and reconstruction is undertaken by the government administration. The government departments are responsbile for constructing and allotting houses to displaced people. Irrigation, PWD and other departments also undertake reconstruction of damaged infrastructure. Development activities in the area affected by disasters are undertaken as part of overall plans. These days, NGOs have come forward to share the burden of the rehabilitation efforts. Some corporate have also undertaken major burden of reconstruction of houses, schools, Public Health Centres (PHCs) and even community centres.

4.11 The entire command, control management and administrative chart of Disaster Management interventions in India is given below. The chart also shows the coopting agencies and their respective level of interventions. It is pertinent to note that during a particular disaster, though the command chains exist, the control for a specific operation is required to be decentralized for a greater synergy and effective management. In India, this aspect is not so evident as yet due to gaps in awareness at various levels and the system not being so vibrant as yet. However, the efforts towards making the system vibrant and responsive have been initiated and in a foreseeable future, India is expected to come up to the required standards. For this, the young generation is required to be made to participate in this field right from the beginning.

DISASTER MANAGEMENT STRUCTURE IN INDIA.

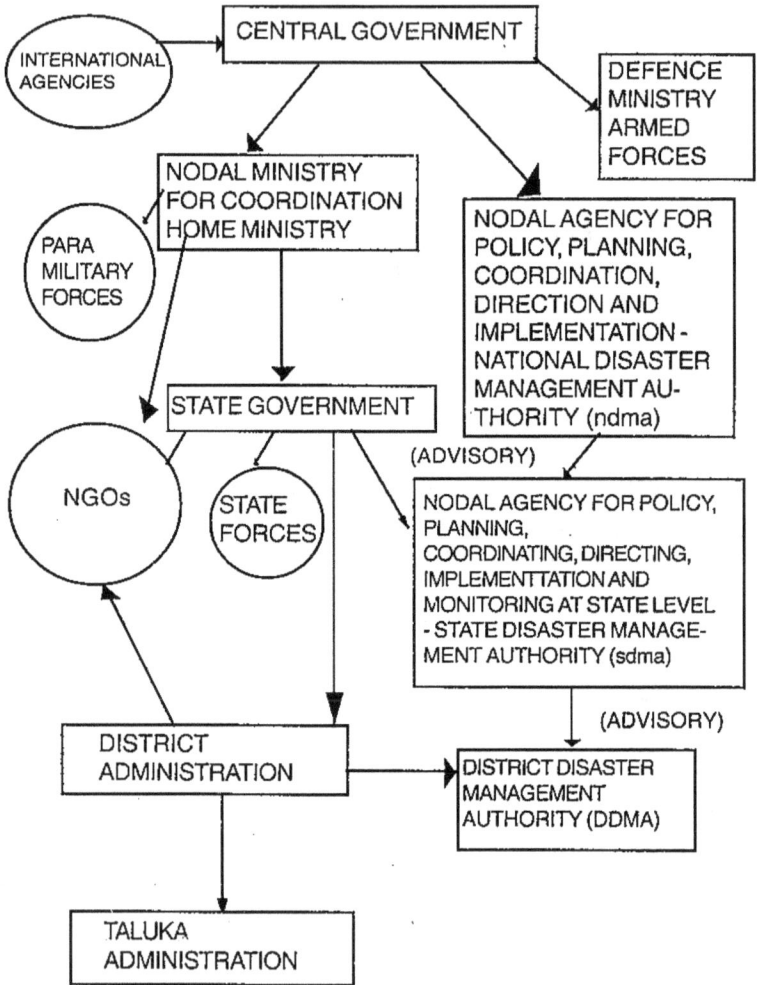

4.12 Other Organizations. When we consider government related organizations, theirs is the main responsibility of mitigation. However, to prevent and minimize losses, institutions, corporate, industries and housing societies must create their own organizations that could be effective within their jurisdictions. Such organizations must have a disaster marshal who should be a senior official of that organization. Under him, a pool of trained volunteers must function at all times and upgrade their preparedness towards mitigation. Such an organization must oversee safety and security of employees, assets and material within the organization at all times and must react speedily if any disaster occurs within or around the premises of such an organization. The outline duties should be :-

4.12.1 Assessing threats and vulnerability of own structures and procedures.

4.12.2 Instituting regular checks.

4.12.3 Safety measures.

4.12.4 Forming teams.

4.12.5 Indentifying safe places.

4.12.6 Working out rescue and evacuation plans under different contingencies.

4.12.7 Identifying place to establish alternate control centre for dealing with critical situations within the organization.

4.12.8 Upgrading training of all the employees.

4.12.9 Holding adequate resources.

4.13 An ideal structure in a corporate could be as under-

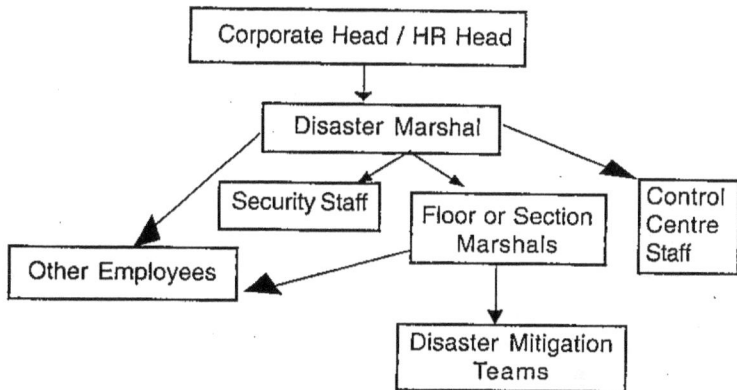

A similar structure could be worked for each institution or organization. Housing societies have to work out an informal structure with dedicated task allotments. Industries may have to adopt a flexible structure as per the shifts.

4.14 A sample organization at Education Institute could be formed out of their present structure as under :-

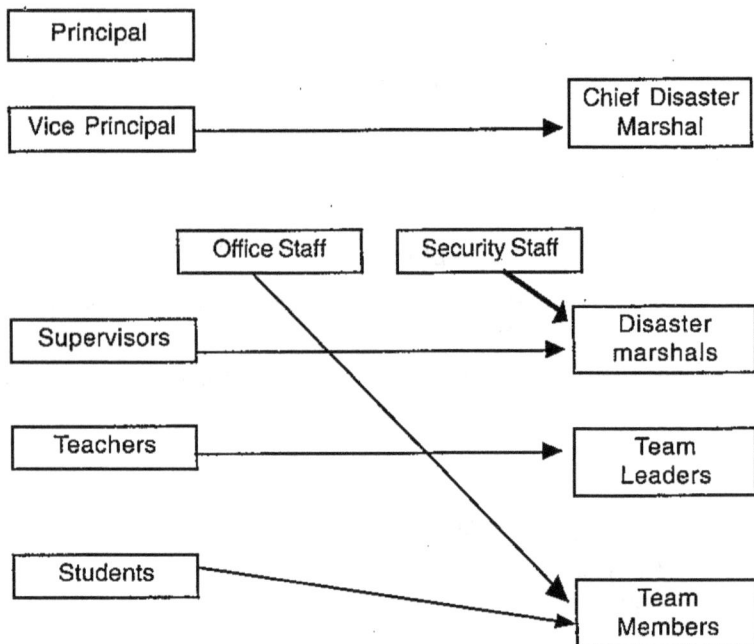

This orgainzation should continously monitor the security and safety of the institution. They should formulate and upgrade mitigation plans including evacuation of inmates to safe places, their administration, fire prevention and fire fighting measures, measures during public disturbances etc. Security staff and senior students should be continuously trained. Office staff, teachers and supervisors must work out the details of administration and also establishment of a control centre and plan to intimate to parents. This aspect should be intimated to the parents also and warning displayed on the notice boards. This is discussed in detail in chapter 7.

Role of Public

4.15 There is no doubt that whenever a disaster occurs without warning, the first few hours are the most grueling ones. The rescue agencies take time to deploy. During this critical period, it is the public that is available at the place of disaster that helps rescuing people and providing immediate relief. In India, most people are not trained in rescue, casualty evacuation and immediate relief duties. Thus, their efforts are unstructured and at times lead to mishandling and ill management of casualties. This is one major reason why every able-bodied person must be trained in the above mentioned aspects and perform simple rescue and relief duties.

EXERCISE

Q1 : How NGOs help in disaster management ?

Q2 : What is the part played by the armed forces in Disaster Management ?

Q3 : What is the responsibility of public during disasters?

Q4 : If you were to initiate forming of an NGO, what structure would you recommend for participating in disaster management related activities at your district level ? Answer in about 200 words.

❖ ❖

CHAPTER 5

RESCUE

Introduction

5.1 : Rescue is the most dynamic and critical aspect of disaster mitigation. It is the most pious of acts towards humanity and is also very ardous, laborious, patience testing, risky and therefore, a challenging job. Rescue tests a rescuer's alertness, compassion, creativity, practical skills, common sense, will power and above all emotional strength. History of disasters is replete with examples of valour and unbelievable will to overcome odds. It has witnessed victims who were rescued after days of sheer grit and application. The most difficult of the rescues have been from fires, mine collapses, accidents in the high altitude regions and even earthquakes. All this indicates that rescue is never a 'chance.' It requires deliberate efforts and a "never-say-die" attitude.

5.2 : When a disaster occurs, inspite of how speedily the public informs any of the rescue agencies, the time lag between such call and the actual commencement of rescue operations by such an agency varies from place to place. The distance between the place of disaster and the location of the agency matters, apart from road communication conditions. During this time lag, the people present on the spot help in every manner to save others. The degree of success of such efforts depends on the skills that people possess individually and collectively.

5.3 : Rescue involves saving people who are trapped and moving them to a safe place. The people who are trapped may be conscious or unconscious. Depending upon the type of disaster, people have to be moved to safety from fire or water, smoke filled rooms, debris of fallen structures or landslides, wreckages of accidents or may be from industrial hazards. The type of disaster dictates the type of injuries or threats and methods of rescue would differ accordingly.

5.4 : Rescue Types

Any rescue can be considered in two main types, the Immediate Rescue and the Deliberate Rescue. They differ in their scope and degree of effectiveness.

5.4.1 Immediate Rescue. The moment a disaster strikes or warning is given, the people who are present in that area rescue the victims from a place of threat to safety. These efforts may not be structured. However, they do help some people to get saved. The immediate rescue shows following characteristics :-

5.4.1.1 Most of the rescuers are not trained and they show courage, after ensuring own safety and help others.

5.4.1.2 Since the people who are involved in immediate rescue are not trained, their efforts are limited and there is a likelihood of their mishandling the victims while rescuing.

5.4.1.3 Right type of equipment may not be available with the rescuers and hence, the effectiveness of their rescue is low.

5.4.1.4 The people undertaking rescue are only in the form of a courageous mob and they lack planning, coordination and a dynamic leadership.

5.4.1.5 However, the efforts are made immediately, without any time delay and thus, help in bringing down casualties.

5.4.1.6 Because there is no centralized direction, immediate rescue may lead to chaotic situation.

5.4.2 Deliberate Rescue. Deliberate Rescue is undertaken by an agency or organization that is especially created, equipped and trained to undertake such rescue tasks. The characteristics of this type of rescue are as under :-

5.4.2.1 The rescue team is trained with centralized leadership and directions.

5.4.2.2 The agency is homogenous and its effectiveness is greater.

5.4.2.3 Since the agency has adequate equipment and experience, their rescue is more efficient and the chances of executing even difficult rescue are good.

5.4.2.4 The effectiveness greatly depends upon the reaction time that lags between the occurrence of a disaster and the actual deployment of the agency.

5.4.2.5 There is a greater coordination between the rescue agency and the government administration and hence, whenever a need is felt it is possible to build up extra efforts from outside.

5.4.2.6 Since the rescue agencies have better communications, transport facilities a vision of the degree of damage and efforts required, the effectiveness enhances manifold.

5.5 Rescue Process

A **Deliberate Rescue** process has many stages. The main stages are - Planning, Preparing and execution. We shall see these in brief.

5.5.1 Planning. The process of planning has certain set procedures and some planning does take place the moment warning is received. The set procedures are called "Standard Operating Procedures." These could be compared with procedures that any surgeon follows while operating upon a patient. Notwithstanding who the patient is, certain procedures and sequences do not change and the need to have some basic equipment remains unaltered. Depending upon the peculiarities of cases, some additional procedures and some specialized equipment may be used by a surgeon. Likewise, an agency undertaking rescue has to have certain pre-set procedures and certain pre-determined equipment availability before a disaster strikes. At the government's level, the planning also involves identification and pre-allotment of certain rescue units and NGOs to undertake rescue operations in particular areas and planning the coordination aspects for smooth execution. The process of planning can be considered in two stages - a pre-set planning and planning for a specific disaster incident. We shall see both these.

5.5.1.1 Pre-set Planning. The pre-set planning has to have the following ingredients -some of which may be considered as preparations.

- Planning a rescource list for each type of disaster and identifying the quantum that would be required for different intensities of disasters.

- Identifying routes and pre-planning movement to different parts of the areas of responsibility.

- Planning the logistic support during execution phase.

- Planning the communications within the organization and with other government agencies.

- Identifying teams comprising of specialist rescuers for different disasters.

- Planning of basic organizational structures for conducting rescue and identifying flexible restructuring based on needs.

- Planning the organization of a control room at disaster site and identifying the government agencies for coordination, in different areas.

- Planning the immediate relief and survival kits and planning to hold them.

- Planning the standard rescue process for different contingencies.

5.5.1.2 Planning When a Disaster Occurs.

When a disaster strikes or a warning for the same is given, the following detailed planning has to be done :-

- Planning the manpower distribution & composition of teams, apart from the initial planning.
- Planning of equipment profile for rescue.
- Planning the actual movement of the teams and coordination with the local authorities.

- Assessment of situation at the site of disaster through a detailed reconnaissance.
- Plan approaches upto the disaster area / site.
- Assess safe and feasible approaches to get inside the site of disaster.
- Establishment of control room / centre at the site.
- Plan on isolating the site from general public and it's safeguarding.
- Plan execution of rescue on the site. For this, the team leaders have to identify structural damages,. safe places, likely places where victims could be found, plan excavation of debris and its removal.

5.5.2 Preparedness.

Preparedness prior to a warning goes hand in hand with the planning. Unless the level of preparedness is high, execution of the actual rescue cannot be effective. Thus, preparation is required mostly much before a disaster strikes. Only a marginal level of preparation needs to be undertaken when actual warning is issued or information of disaster is received. Once the information about a disaster is received, the execution stage gets started. Let us see what the preparation should be :-

OPERATIONAL PLANNING

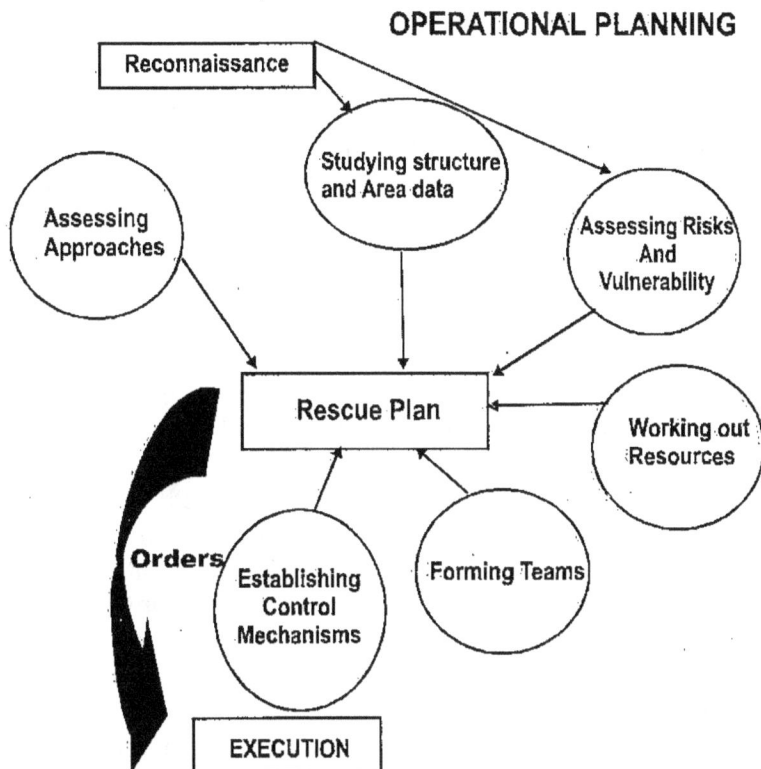

5.5.2.1 Organization. Creating an organization to conduct rescue and relief operations is the basic preparation that any country requires. Only an organization that is tailor-made for this purpose could serve the right ends. Armed forces are inherently geared up to face any criticality and thus, they are most suitable for this purpose. However, in performing such tasks, they are diverted from their main tasks related to defence. Thus, there is a need to have an organization, especially created to deal with disasters and thereby bringing down the burden on the armed forces. While creating such an organization and locating it suitably, a detailed thought has to be given to the aspects of manpower, equipment profile, training and skills imparted and the pre-location depending upon the threat analyses. Location has to be worked out in a manner that the reaction time from warning till the actual deployment of the force is minimal. Moving and deploying such a force forward during warning period is also a part of anticipatory preparedness.

5.5.2.2 Command and Control Structure. For smooth passage of warning, deployment of rescue efforts and dynamic feedback, a command and control structure has to be created. The efforts of the central government to create organizations at the Central and State Government levels and also at district levels and identification of rescue force and pre-locating it is an effort towards this end. The principle that brings the best results is of 'Centralized Command and Decentralized Control. It simply means that the planning, coordinating and directing of efforts is done at national level and the actual execution is done at state and district levels. Execution can only be smooth provided the command structure to deal with the disasters is in place and the control mechanisms are created. Thus, a national level control centre, state control centres and district control centres have to be created well in time. These control centres have to be adequately staffed, equipped and put into various captive communication channels. The functioning during execution phase can be smoother if such control centres are well orgainzed. A sample control centre at district level has been shown in the following diagram.

Security Fence

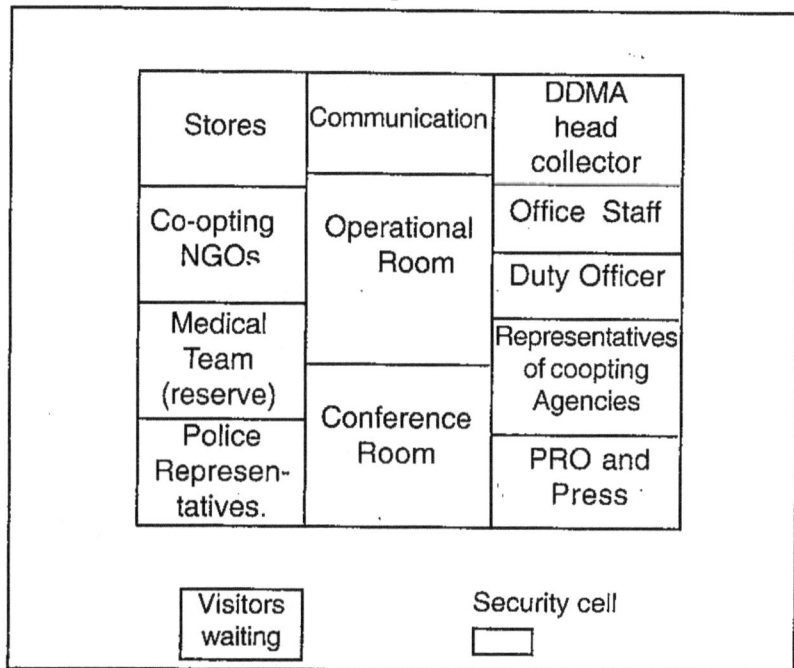

Stores	Communication	DDMA head collector
Co-opting NGOs	Operational Room	Office Staff
		Duty Officer
Medical Team (reserve)		Representatives of coopting Agencies
Police Represen- tatives.	Conference Room	PRO and Press

Visitors waiting

Security cell

5.5.2.3 Information Collection. During the preparatory stage that runs simultaneously with the pre-disaster and execution phase, information collection becomes one important aspect towards preparedness. Firstly, information collection channels are required to be created by the government. For this, the most important is the selection and deployment of warning systems. Next is a huge static data regarding threat, vulnerability and risk analyses. The information should be of geographic, geological, demographic and economic nature. This information helps in preparing further for executing rescue. All this information may fall into static and dynamic fields. Information collection and dissemination procedures must be worked out in the preparatory phase.

5.5.2.4 Preparedness at Rescue Unit Level. Preparations to face any disaster are a continuous process. A rescue unit should perform the following preparedness acts during pre-disaster phase and after warning or information of any occurrence of disaster is received. We shall see this in two parts :-

5.5.2.4.1 During Pre-Disaster Phase. The following preparations are expected :-

- Working out equipment loads and maintaining the equipment in "an all time ready" state.
- Rehearsing various rescue drills as mock practices.
- Acquiring greater knolwedge and skills.
- Rehearsing movement plans to different areas within own jurisdiction of responsibility and establishing coordination with the government administration at the District Level, Taluka Level and with other similar co-opting agencies.
- Practicing communications and also control room operations.
- Updating all maps and information.

5.5.2.4.2 During Active Phase.

The following preparations are expected :-
- Loading and re-formation of teams based on the information.
- Conducting advance reconnaissance to assess the situation in coordination with the district administration and the local people.
- Making dynamic outline plan of rescue and briefing the teams.
- Loading the transport and working out logistics.

Execution of Rescue

5.5.3.1 Execution is conducted in the following sub-stages:-

Movement of the rescue unit / teams to the area of disaster.

5.5.3.2 A control room or centre is established at a suitable place where the room is not affected by the disaster.

5.5.3.3 Cordoning off the area for security reasons.

5.5.3.4 Once deployment plan for the rescue is finalized, the teams are reformed and briefed in detail. Coordination between teams is done if there are more number of teams operating in the area.

5.5.3.5 Launching the teams for Rescue. When teams are launched, depending upon the type of disaster and its magnitude and the condition of the site, the teams approach and enter the affected area. The rescuers operate in minimum of pairs. The rescuers locate the trapped victims under debris, fire, smoke or even possible places of drowning. All structures are searched. During floods, poles and piquets are firmly anchored and ropes and inflated tubes are tied across the water channels. Boats are launched for search and rescue. Marooned people are moved to safety. Digging of the debris is conducted carefully. The rescuers use every possible material to recover as many victims as possible.

5.5.3.6 The victims that are located may be in conscious or unconscious and injured state. These victims are stabilized within the affected area and then evacuated to a safer place outside the affected area. Immediate first air and resuscitation may be offered. The rescue teams may take help of some courageous local people for rescucitation and carriage of casualties.

5.5.3.7 After rescuing possible victims, the rescue teams also try to salvage property which is then guarded and handed over to the police.

5.5.3.8 Rescue teams also recover dead bodies which are handed over to the police and the effects on the dead bodies are listed and handed over.

5.5.3.9 The entire process of rescue is totally situation dependent. The following diagram shows the execution process.

Diagram showing Execution of Rescue Process

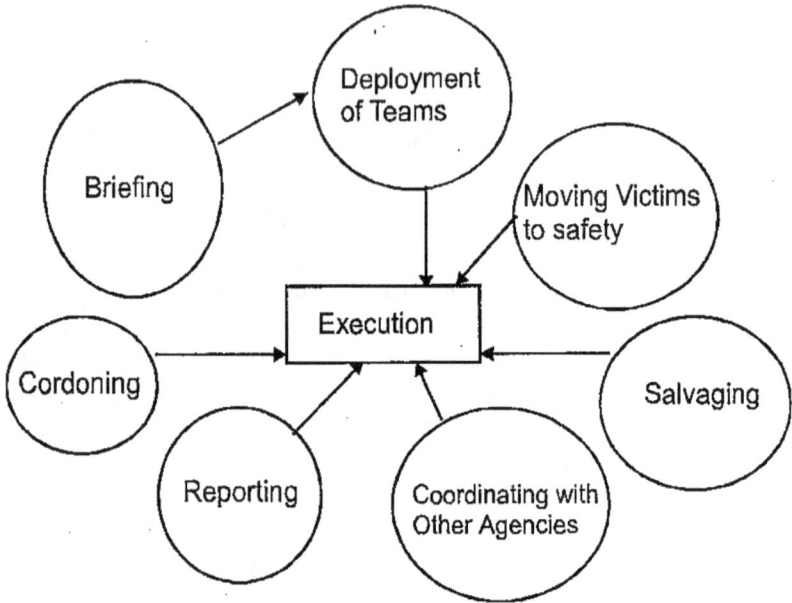

5.5.3.10 The rescue unit reports progress of rescue to the coordinating government agency (District Collector or Talukha Office). The orders to discontinue rescue efforts are given by the District Collector, depending on the situation.

An after action report is rendered by the rescue unit (agency) to the goverment administration, apart from rendering a similar report therough own command and control channels. These after action reports are very important from the viewpoint of future policies and correcting past anomalies.

5.4 Rescue Process Training for General Public

A common man can join an NGO involved in Rescue and Relief opertions and get trained. A common man can get trained even to facilitate ownself to conduct immediate rescue whenever opportunity exists. But, for that a particular type of basic training is required. It is the responsibility of every citizen to rescue own self. The training would go a long way in achieving this. Following paragraphs list out the type of training that every adult is expected to take. It also gives a list of training that rescue organizations must undertake.

TRAINING ASPECT	DETAILS	TIME FRAME	IMPARTED BY
Disaster Awareness	• Causes, effects and characteristics of different disasters· • General Preventive measures • DO's and DON'Ts	3 hours	NGOs, Civil Defence and Fire Brigade.
Warning systems	• Different warning systems for different disasters· • Actions on receipt of warning	1 hour	NGOs, Government Agencies, Civil Defence, Police and Fire Brigade.
Self-Rescue	• Finding safe Places when trapped in a disaster and moving out of unsafe areas· • Survival techniques while in safe places — securing own self, surveying area around, making bigger space for breathing, avoidance of perspiration and surviving without food and water for long durations. • Sending out messages about own location. • Use of some implements of daily use.	3 hours	NGOs, Civil Defence, Fire Brigade

TRAINING ASPECT	DETAILS	TIME FRAME	IMPARTED BY
Rescuing own self and others with Innovations	• Moving out of fire and fire fighting • Moving out from a smoke filled area • Techniques related to floating in water for swimmers & nons wimmers, creating floats and rafts • Breathing control techniques • Moving out of Landslide area and self protection from falling debris • Using ropes and knots in rescue • Taking out casualty from an accidented vehicle or a train	4 hours of theory and practice (to be revised on regular basis)	NGOs, Civil Defence and Fire Brigade.
First Aid	• Self Help • ABC of first aid • Artificial respiration • Dealing with wounds & fractures Evacuating a casualty	4 hours	NGOs, Civil Defence, First Aid Training Organizations (like St John Ambulance)

Total time required is 15 hours. This could be spread over 2 to 3 days. Such courses should be regularly run at educational institutions, housing colonies and offices.

5.5 Training that is expected to be taken by other organizations (Training at housing Society, Institutions and Offices) is given in the following table :--

	DETAILS	TIME FRAME	IMPARTED BY
	• Safe exits in a particular building complex • Use of Fire Fighting equipment in tandem and other means • Clearing debris to reach trapped victims without use of machinery • Using innovative methods for a collective rescue	3 hours	NGOs, Civil Defence and Fire Brigade
	Use of multiple manpower	1 hour	NGOs and Civil Defence
	Organization of an adhoc control center and forming	2 hours	NGOs and Civil Defence

TRAINING ASPECT	DETAILS	TIME FRAME	IMPARTED BY
	of command structure quickly to control community actions & sending out waning to rescue agencies		NGOs and Civil Defence
Relief	• Arranging accommodation and holding inmates • Arranging administration of inmates and their safety and security • Informing family	2 hours	
Making Mitigation plans for the specific complex	• Pre-deciding disaster mitigation teams and earmarking their duties • Decide on communications and safe places for evacuating victims and administering them, planning options for exits and instructing the entire staff, students and laying down SOPs	This is a deliberate process and has to begone into by an institutional head with the help of experts in the field. This may require a few days or weeks to implement. It should also include vulnerability analysis and equipment required to mitigate and its placement and use.	NGOs, Civil Defence, Fire Brigade and other experts.

A total of 8 hours would be required. Teams should be trained in Module 1 & 2 at village level and Block levels, under the administration of BDOs and Village Panchayat.

Training that is essential for general rescuers is given in the following table -

5.6 General Rescue for Trained Rescuers

TRAINING ASPECT	DETAILS	TIME FRAME	
Rescue from Debris	• Planning the rescue • Creating Access • Clearing Debris manually and with tools and machinery • Securing trapped victims • Evacuating victims to safety • Control Center operations and coordination of teams	2 hours of theory and 8 hours of practical.	
Rescue from water	• Creating safety nets •Operations of boats •Creating rafts • First aid to a drowned victim • Evacuating endangered people to safety with the help of poles, bamboos and ropes • Recovering a drowning victim	2 hours of theory and 8 hours of practical	

TRAINING ASPECT	DETAILS	TIME FRAME	IMPARTED BY
Rescue from Fire and Fire Fighting	• Fire fighting and smoke drills • Carriage of casualties out of fire • First aid to fire victims	8 hours of theory and 8 hours of practical.	Fire Brigade, NGOs, Civil Defence
Rescue from landslides and mountainous terrain	• Basic mountaineering skills to also include rappelling and rock climbing • Survival techniques	Basic mountaineering coursrewith casually evacuation 8 hours.	Mountaineering institutes like HMI, will have to be specially designed for rescuers

A total of 44 hours of training is required for Module 3. Before a person is trained through this module, it is important for the person to go through Modules 1 & 2.

5.7 For Government Administrators.

This module should include the administrative aspects and organization of rescue and relief work.

TRAINING ASPECT	DETAILS	TIME FRAME	IMPARTED BY
Disaster General	• Causes, Effects and characteristics of all types of disasters and their prevention • Multi-disaster structured and unstructured responses	4 hours	Government Administrative Staff Colleges with Resource persons from the community
Organizational Issues	• Organizing Controlling bodies and mechanisms	8 hours	-DO-

TRAINING ASPECT	DETAILS	TIME FRAME	IMPARTED BY
	• Organizing Resources for Rescue and Relief • Coordination with other agencies • Establishment of control centers and communications • Organizing and administering relief camps		
Survey Aspects	•Pre-Disaster surveys and post disaster surveys.· ·Organizing survey teams· •Parameters for damage assessment.	4 hours	-DO-
Risk and Vulnerability Analysis of a District and a Tehsil	• Theory of threat, vulnerability and risk assessment· Preventive measures	2 hours	-DO-
Planning mitigation	• Making Mitigation plans and working out resource mobilization and deployment· Coordination of processes from International to Tehsil Level • Warning systems and their deployment· Creation of 'Self-Help' groups from the community •NGO's participation and control	4 hours	-DO-

TRAINING ASPECT	DETAILS	TIME FRAME	IMPARTED BY
Field Survey	• Visiting a threatened area or an area of an earlier disaster and conducting survey		-DO-
Simulation Exercises	• Simulation exercises to cover all phases of disasters on a simulated area, with all agencies co-opted		-DO-
Group Discussions & Experience Sharing	• Discussing various modalities of execution during a disaster		-DO-
Legal Aspects of Disaster Mitigation	• Laws prevalent and implications of various situations		-DO-
Aspects of compensation, Rehabilitation and Re-construction	• Practical aspects and procedures		-DO-

5.8 In India, we pay scant attention to the training aspects. This anomaly needs to be corrected by making the basic practical training obligatory. They say, "the more we sweat during peace, the less we bleed during wars." This is highly applicable to disasters as well; for we the humans, to wage a war against the disasters.

EXERCISE

Q1 : What is a rescue ?

Q2 : Which agencies can be employed for rescue ?

Q3 : What preparations should a rescue unit do during pre-disaster phase ?

Q4 : How is rescue actually conducted ?

Q5 : What preparations would you advise when warning for flood is received ?

Q6 : What are the essential equipments that a rescue team must have while rescuing people from :

(a) Floods.   (b) Fire. (c) Building Collaps.

Q7: How would you organize training concerning Disaster Management for your college / institution?

❖ ❖

CHAPTER 6

RELIEF

6.1 : Disaster throws up multifaceted effects. These effects range from superficial injuries to ghastly deaths, partial damage of property to complete loss, washing out of crops to spoiling of material due to accumulation of silt and debris. All these effects warrant and elicit actions from the government and community towards restoration of normalcy. Relief could be defined as all actions - structured or unstructured - taken to restore normalcy to life. It may encompass provisioning of material for immediate sustenance, offering of shelters and helping in rehabilitation of the victims. In some cases, the relief may extend towards reconstruction as well. Like the rescue acts, relief also needs to be planned and executed according to some norms. We shall make efforts to study the aspects of relief, to bring all actions within a structured framework so that relief could be more gainful and logical.

Relief as a Process

6.2 Relief is also a process that commences either when warning is received and people are moved to safe places or even immediately following the commencement of rescue process when the trapped victims are moved to a safer place and have to be offered immediate sustenance. For an effective relief, the process has to be triggered during preparatory phase. Like the rescue, relief process also has two major stages - Planning Stage and Execution Stage. Relief also is of two types - immediate Relief and Deliberate Relief. The immediate relief has two further parts - incidental relief provided on the spot by people who are present at the disaster site and the pre-planned immediate relief that is provided by the government and other co-opting agencies. The following diagram depicts the whole process of relief :-

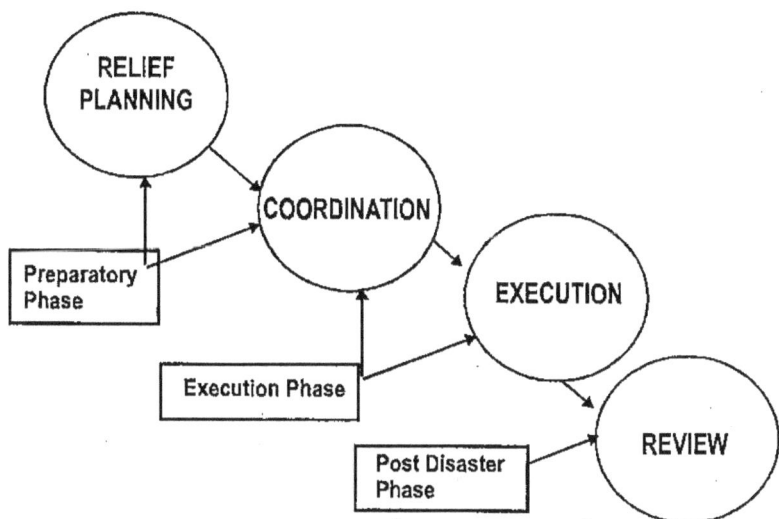

Preparatory Phase of Relief

6.2.1 Preparatory Phase of Relief. This phase is part of an overall Preparatory Phase of Disaster management. This phase includes anticipation of effects of any disaster through Threat, Vulnerability and Risk Analysis. Such an analysis helps in estimating the number of people likely to be affected. Once this estimate is obtained, the process of planning gets triggered whereby the provisioning action commences. For different types of disasters, the requirement of relief material differs. For example, for floods, shelters, food and clothing along with medicines may be prime requirement; whereas, during draughts, water and food could be the prime wants. The relief agencies - government and NGOs need to keep this aspect in mind. In our country, the relief material is catered for only after the requirement is felt. However, a pro-active relief plan could entail much more than that. Relief could also be seen as immediate relief and deliberate relief. The general public provides the initial part of immediate relief on the spot in whatever terms possible. However, the planned immediate relief is the basic responsibility of the government and the co-opting agencies (NGOs). The preparatory phase should include the following :-

6.2.1.1 Resource Planning for Immediate Relief. The government and the co-opting NGOs must plan on modular system of resource requirements based on a base line of 500 victims or so. Thus, a list of items and quantum that is required for immediate relief should be worked out. This should include the following :-

Two sets of clothing per person - catering to 40% adult males, 40% adult females and 20% children.

Sheets for under - spreads and covering.

Food packets per day of cooked food and approximately 2 days of rations to be served after cooking in situ, through community kitchens.

Clean drinking water at the hard scale of 5 litres per person and water for utility purposes at the scale of 20 litrs per person.

Shelters (tentage) at the scale of 48 tents for a relief camp. (for 500 people)

Suitable quantum of medicines.

Out of the above requirements, the perishable goods have to be managed on as required basis while non-perishable goods could be catered for through permanent stock holdings.

6.2.1.2 Resource Planning For Deliberate Relief. Deliberate relief includes the following :-

Provision of temporary shelters deliberately constructed and suitably located.

Provision of means to run family kitchens (utensils, stoves and dry rations etc).

Provision of material for restoring normalcy on life (like educational aid, supply of means of livelihood etc.) To give an example, after the Tsunami, the government and other agencies offered repair of fishing boats and also provision of new boats to the fishermen who had lost their boats. This aspect is difficult to plan prior. The requirements are worked out only after immediate relief is provided, based on actual figures, rather than estimates.

6.2.1.3 Stocking of Relief Material and Its Accounting. The system of pre-stocking of relief material is presently non-existing in India. However, such a system is highly desirable. As brought out earlier, the stocking of basic commodities for speedy relief is essential. Whenever any material is held captive, its stocking and accounting pose problems. There is a fixed shelf life for every material and possibilities of decay do exist. Thus, captive stocking has to be coupled with maintenance of the stocks and detailed accounting. It is but natural that for performing such tasks stocking facilities have to be created and also a dedicated manpower has to be employed.

6.2.2 Coordination. The preparations also include coordination with other agencies like the Military and Para-military forces, NGOs, Corporate and other resource manufacturing and dealing agencies. Many times, it is not possible to pre-stock materials to required extent as it becomes uneconomical. In such cases, the resources could be created through coordination. The coordination would involve the following :-

6.2.2.1 Coordination of Medical Efforts. This involves listing of doctors, paramedics, equipment and medicines to be made available through coordinated efforts of all agencies. Many times hospitals have to be earmarked for increased bed capacity. Private specialists also have to be approached for contributing to the efforts. In case of L3 level disasters, apart from mobile teams of doctors pooled in by the government, the teams from armed forces, para-military and NGOs have to be deployed. This has to be thought about during the preparatory phase itself and prior coordination has to be done. Agencies have to be identified for supply of large quantum of medicines. During floods in Mumbai in July 2006, there was a requirement of large quantum of "Doxycycline" to fight Leptospirosis. This was in short supply in India and had to be flown in as an emergent measure from other countries. This may be a special case. But, medicines for water purification, general anti-biotic, anti-fungal and many other medicines that are required in large quantum may be provisioned through distributors and the industry.

6.2.2.2 Coordination for Other Supplies. The other supplies include food material that may be perishable. Such material like fresh vegetables, milk etc has to be supplied through contracting and the sources have to be tapped well in time. Non-perishable material could be preserved as explained earlier. On the first two days while the camps and cooking facilities are not ready, 'ready -to-eat' food packets have to be distributed with bottled water. The sources making available such food packets have to be well identified. Bharatiya Jain Sanghatana is one such organization which has been preparing healthy food packages for airdrop and land based distribution to disaster victims.

6.2.2.3 Clothing. The victims require clothing. In different areas, the needs differ. Also, the government has to cater to the needs based on gender composition and an additional number has to be catered.

6.2.2.4 Coordination of Camp Sites and Organization. The relief coordination also involves identification of places for camp siting. As an immediate measure, schools and colleges are used by the government. However, this could only serve the purpose for a limited time. For a longer time period and to ensure good hygiene, separate camps have to be set up. A camp for 500 victims and the administrators needs an area of aproximately 150 mtr x 150 mtr. These are practical planning figures. In most towns, such available areas are few as most of the land is in possession of private owners. The coordination aspects include earmarking land and coordinating with the private landowners against reasonable compensation. This aspect involves legalities and adequate provisions need to be done in the country's legal framework.

6.2.2.5 Coordination for Distribution and Accounting. A major problem that has been observed during disasters is that of the government's ability to reach inaccessible areas and distribute the essential goods and services. Most of the time, some areas are covered in excess while some areas remain unattended. The government does not have adequate manpower to conduct actual relief work. Since 2005, attention is being paid towards this aspect. Help of NGOs should be taken and responsibilities should be distributed through an in-depth coordination during preparatory phase only.

6.2.2.6 Coordination of Survey Efforts. After a disaster strikes, it is difficult to assess exact extent of damage. This happens because; the affected area is not always accessible immediately. The government has to plan and coordinate collection of information through pre-earmarked survey teams. The survey information has to be used for distribution of compensation, planning and execution of rehabilitation and re-construction.

Execution Phase of Relief

6.3 The execution Phase includes dynamic Relief activities. The relief activities during this phase are conducted almost concurrently with the rescue activities. A rescue team has the sole responsibility of moving disaster victims to safety and whenever required, stabilize them at the point of threat before making them suitable for evacuation. Such victims are then handed over to the 'Relief Team'. For example, the rescue team may find trapped victims who are in need of immediate first aid and resuscitation before attempting their evacuation. For this purpose, a rescue team is also well equipped. However, the teams cannot operate in watertight compartments and therefore, if required, medical aid from relief team, if necessitated, is sought. After the casualties are stabilized and are rendered fit for further evacuation, they are evacuated to the waiting hands of the relief team members. Rescue acts involve the following in execution phase of disaster mitigation.

6.3.1 Deployment of Different Relief Teams. Relief teams should comprise of the following elements :-

6.3.1.1 Establishing a Victims Collection Point at the Disaster Site. This is established in the form of a reporting centre where victim of any category reports for help. Here, the organization should record the detail of the victims and render immediate help in terms of food and water, medical aid and if required emergency clothing. The victims are held only for a short period and are evacuated to a relief camp.

6.3.1.2 Evacuation of Victims to Safer Places and Relief Camps. During warning period or after a disaster strikes, evacuation of the probable victims to safer places becomes a massive operation when large number of people are affected. Transportation efforts have to be worked out and routing has to be planned. The government is expected to keep homogeneity of the victims in such a way that the families are not split and also people from the same villages or localities are kept together as far as possible. If this is not assured, there is a likelihood of chaos and some victims may suffer. During Hurricane Katrina, the administration of New Orleans failed to evacuate the population in time and people got stranded. This happened, inspite of a three-day warning period available.

6.3.1.3 Establishment of a Relief Camp. To administer a relief camp well, following organization is suggested :

6.3.1.3.1 A Camp Commandant. He should be from the organization establishing the camp. He should be overall responsible for the entire administration and security of the camp and ensuring that the inmates are given the minimum basic acceptable material and comforts. His decisions and actions help restore early normalcy in the lives of the inmates. He should control the entire activities of the camp and act as the "Responsibility Centre".

6.3.1.3.2 A Camp Administrative Coordinator. This person should be from the district or Tehsil administration and would be the liasion between the Camp Commandant and the Government. Since such a person would be from the local administrative body of the government, he / she would be in a better position to interact with the victims regarding the government's decisions and plans and understand the difficulties of the people. Many a time, the camp commandant may not know the local language and hence, the coordinator is essential.

6.3.1.3.3 A Camp Staff Officer. He / She should be an experienced volunteer and preferably a little senior in age. He / She assists the Camp Commandant in coordinating all administrative functions of the camp.

6.3.1.3.4 A Team of Volunteers. Volunteers are required to look into the food, water, sheltering requirements and any other provisions distributed in the camp. The volunteers are also required to look after hygiene and sanitation requirements and security of the camp. The same volunteers also look into maintenance of various records, store keeping and even the entertainment aspects. For this to happen, a camp commandant needs to make various teams from the volunteers and execute the administration. Strength of this team depends upon the strength of the inmates. It has been experienced that for a camp that accommodates 500 victims, initial setting up efforts are intense and a team of 40 volunteers is adequate to perform the setting-up duties and initiate the administration. After the camp settles down in the first two weeks, a team of 12 to 15 volunteers is good enough to continue the administration if the inmates are also inducted to perform certain functions like cooking and hygiene and sanitation related duties. In fact, involvement of the inmates is highly desirable because it gives them a sense of belonging and achievement and helps resolve many administrative and psychological issues.

6.3.1.3.5 Medical Team. Every camp should have a dedicated medical team of doctors, paramedics and clinical psychologists. The strength of this team would depend upon number of inmates and proximity of two or more camps. While it is desirable to have a dedicated team for each camp for the first few days, reduction in the strength or combining of efforts could be resorted to after the victims' conditions start improving and the victims settle down. However, experience during earthquake in Bhuj (Gujarat) and tsunami in Tamil Nadu and Kerala suggests that for every 500 inmates, a team of three doctors, three to four clinical psychologists and two to three paramedics is sufficient for the first one to two weeks when the casualty and ailment rates are high. Thereafter, one doctor and one paramedic and one clinical psychologist could take on the routine burden for the next one month. After that, the medical team could be deployed on need basis.

6.3.1.3.6 A camp must have enough space to accommodate the inmates. A space of 150 metrs X 150 mtr is found to be just adequate for a camp for 500 inmates. There should be separate tents for running a community kitchen, ration stores, medical stores, a medical treatment centre, a report centre and for the volunteers' living and administration. There should be enough material and space to establish toilets. A camp should have enough provision of water, security lights and even some transport for administrative duties. We shall discuss the camp establishment and administration in more details.

6.3.1.3.7 Administrative Aspects of a Relief Camp. The following points require due consideration :-

6.3.1.3.7.1 Infrastructure. The points brought out hereunder are based on previous experience in India. A similar template could be used for other developing and under-developed countries. For a camp to be effective, the following conditions should be met :-

6.3.1.3.7.2 Accommodation. A standard tent generally accommodates 15 people. During the initial stages, immediately after a disaster strikes, it is more convenient to accommodate males and females separately, instead of establishing family cohesiveness. Alternatively, a norm of accommodating two to three families in one tent could be followed. After the first effect of disaster wanes away and some inmates start going back to their houses, the composition of victims in the tents could be suitably altered to bring each family together and reallocate tents accordingly. This helps to restore the semblance of family togetherness and helps the victims psyschologically. In the initial stages, 35 tents (with capacity of 15 each) are sufficient for accommodating 500 inmates. Over and above this number, volunteers require three tents for their living, two tents are required for the voluntary organization to establish their cookhouse and ration store and three tents are required for the same purpose for the inmates. One tents is required for establishing a report and control center and two tents for medical stores and treatment centre. Two tents are required for storing relief material. Thus, a minimum of 48 medium level tents are required to establish a reasonably comfortable camp. Over and above this, a large number of tent material or plastic sheets are required to establish latrines and bathrooms and also to protect the tents from wind conditions. During rains, the tents have to be also covered with plastic waterproof material. When 500 inmates are required to be readjusted keeping family cohesion in mind, the basis of 5 inmates per tent should be the standard norm. Initially, a community kitchen could be run in each camp. When sufficient relief material is available and when the victims settle down, a stage may be reached when individual family kitchens could be allowed. Such a decision would be situation dependent.

6.3.1.3.7.3 Other Equipment. Apart from digging tools and requisite quantity of material like ropes and pins for tent erection, some important equipment that is required includes utensils and stoves for cooking, water storage facilities for drinking and general usage at the scale of 5 litres per head per day for drinking and cooking and an addition of 10 litres per head per day for general usage. The water availability could be slowly increased with a resupply managed twice a day such that at least 40 liters of water is available to each individual on each day. Also, fire-fighting equipment is required to be stored with fire extinguishers, billhooks and beaters and buckets. Additional ropes are required for establishing security and traffic control. Enough emergency lighting facilities should be built up and hygiene and sanitation chemicals (disinfectants) have to be arranged. Telephone or radio facilities need to be provided to the camp. If barbed wire fencing is provided, it could ensure restricted access and the security of the inmates could be taken care of.

6.3.1.3.7.4 Registration of Inmates. A camp must start accepting inmates after the basic infrastructure is made available. The inmates should be directed by the administrative coordinator to the camps. Every inmate should be registered in a master record, with name, earlier address, date of birth, name of the next of kin and the thumb impression. An identity card should be issued to the inmates for identity. It is worth noting that the inmates generally would have lost all identity papers and this identity card is the only document to recognize the person's existence in case of the relief camp. This identity card also helps in entry restriction later, for security purpose. The identity cards could be for multiple purpose where records of immunization and the relief material distributed could be endorsed. This helps in accounting and avoid duplication.

6.3.1.3.7.5 Distribution of Basic Amenities. Once an inmate is registered into a camp, basic amenities could be immediately handed over and accounted for. The basic amenities for immediate sustenance would include a mat, a blanket, a pair of clothes, a plate, a mug, toiletries and a small water bucket. This should sustain a victim for a few days. Relief could be built up further, later.

6.3.1.3.7.6 Food Distribution. Inmates should be provided with food or snack immediately on arrival in the camp. This helps the victims to relax, psychologically stabilize, feel secured and mix around with the other inmates. Inmates who are psychologically balanced should be tasked for running and administering the community kitchen. While providing rations, the administration and the NGOs must keep in mind basic food habits of people in the region.

6.3.1.3.7.7 Medical Treatment and Immunization. An inmate should be medically examined and immediately treated on arrival at the camp. During disasters, there is a high risk of communicable and water borne diseases. To avoid any risk, every inmate should be immunized for typhoid, cholera and preferably hepatitis. Even the volunteers and medical teams should also be immunized. Wounds and fractures, fever, cough and cold, diarrhea and dysentery, asthma and tensions are most commonly seen ailments. Apart from that, some old people also have chronic ailments and acute conditions of heart, liver and diabetes. Duly formatted case papers would help. Pre-coordinated tie-ups with hospitals nearby could help deal with such cases in absence of previous history and non-availability of advanced equipment in the camp. Medical teams should be accordingly equipped. During floods, tsunami and earthequakes, snakebite cases in rural areas are rampant. A camp should be adequately equipped to handle such cases. Volunteers should dig snake trenches around the camp and educate the inmates about snakebites.

6.3.1.3.7.8 Hygiene and Sanitation. This is a very important aspect of camp administration. Proper hygiene and sanitation prevents onset of epidemics. The points that need attention are as follows :-

Latrines and bathrooms that are created do not have proper drainage systems as these are of temporary nature. Hence, spraying of disinfectants should be resorted to twice a day.

Community kitchens should be fly-proofed using cotton nets.

Inmates should be given mosquito repellants.

Soak-pits should be adequately dug and disposal of the waste should be done with due care. Inmates should be educated on cleanliness of the camp and teams could be formed from the inmates to frequently clean up the entire area.

6.3.1.3.7.9 Psychological aspects. Disasters create a lot of psychological problems amongst the inmates as well as the volunteers. in the initial stages of a disaster, invariably, many psychological disorders are seen in the victims. The trauma of loosing near and dear ones, the destruction and havoc experienced and fear of the unknown future grip the victims- small children and the elders, alike. Clinical psychologists have a great role to play during this time. The psychologists should prepare formatted case sheets in dealing with various cases. free mixing of volunteers with the inimates and motivational talks help. various philosophical discourses and ' Bhajans' have proved effective during such times. A victim overcomes his/her grief in company of other victims and the strengthening is faster when a victim is kept occupied in community activities. Such creative and constructive activities should be encouraged. The camp commandant must frequentely interact with the inimates and address them together. Where there is a language barrier, an interpreter accompanying the volunteers or the one found from within the inimates should be made use of. Victims often get angry out of sheer frustration. Volunteers should be educated to keep their cool and not have confrontation with the inimates. One must understand that such anger is actually due to the suffrage and it is a way of letting out emotions. Compassion should be the key word in dealing with the inimates. Involvement of the inimates in camp administration and allocation of duties should be actively pursued on the basis of self-governanace. It has been found to be very effective and gives a great sense of fulfillment to the inimates and helps restore normalcy.

6.3.1.3.7.10 Entertainment and sports. Entertainment and sports have a great impact on the lives of the victims. Facilities for entertainment and sports should be created, especially for children.

This goes a long way in psychological build up. Competitions for children and the grown- ups should be organized. Competitions like drawing and painting and music allow free expression to the victims and help to overcome the stress of a disaster.

6.3.1.3.7.11 Records. A camp should maintain various records. These include- Records of inmates, Records of relief material distributed, Records of rations received and consumed, records of immunization and medical records of treatment. Apart from this, a daily event diary needs to be maintained. Many VIPs visit relief camps. these visits should be recorded separately and signatures of the VIPs and the summary of their talks should be recorded and preferably should be got signed by their secretaries. Records as these reveal a lot of information.

6.3.1.3.7.12 Report and Control center. This is the nerve center of a relief camp. it is here that the activities of directing, coordinating and controlling take place. This control center is different from a control center organised by the state and district or Tehsil Authorities. The report and control center of a camp is limited to administering the camp only. It is here that the inmates first report, it is here that all activities and happenings are recorded. A camp commandant establishes his power through this center. It is here that he briefs his volunteers and meets the administrative requirements of the camp. All VIPs and media persons report here for further interactions. During relief stage, a lot of government officials need to interact with the inmates for compensation disbursal, assessing records etc. These activities are conducted at the report and control center. Many outside agencies contact relief camps for distribution of various relief related material. It is at this place that the activities are conducted. A report center also acts as a communication hub.

EXERCISES

Q1 : What is difference between 'Immediate Relief' and 'Deliberate Relief' ?

Q2 : If you are present at an accident site, what relief would you offer to the victim immediately, before taking the victim to a hospital ?

Q 3: What is importance of sports in a relief camp ?

Q4 : What is the importance of hygiene and sanitation in a relief camp ?

Q5 : How would you organize a relief camp and what teams will you form to accommodate and administer 500 victims of an earthquake ?

Q6 : What assistance will require from the district or Taluka authorities ?

Q7 : What documents would you like to generate in the above relief camp ?

Q8 : How will you ensure disposal of sewage, garbage and purification of drinking water ? Explain in about 300 words.

❖ ❖

CHAPTER 7

SOME DISASTER RELATED
INTERVENTIONS

DISASTER PREPAREDNESS OF SOME PLACES
OF IMPORTANCE

SAFETY AND SECURITY AT EDUCATION INSTITUTIONS

Introduction

7.1 Indian culture, over the centuries, has been molded into a state of complacency. It is our tolerance and a bent of mind to accept things as they appear make us more reactive in nature. However, the recent incidents of disasters have set us thinking about safety and security. Educational Institutions in India are soft targets and very vulnerable to disasters - natural and man-made. Fire in a school at Kumbhakonam killed 93 students. A fire engulfing a school "shamiyana" killed 425 students at Mandi Dabwali in Haryana in 1995 and a school bus was over-run by a train at an unmanned crossing near Saswad in Pune District. During the earthquake of Gujrat in 2001, 31 teachers and 971 students were killed. These are some examples of the vulnerability of this section of the society. All these incidents indicate a lack of preparedness, awareness and absence of procedures to safeguard the inmates of an educational institution. It is certainly not enough to impart good education; it is equally important to safeguard the students so that they are able to use the education in future. Though the government has made 'Disaster Management' a part of the syllabi, the institutions seem to be looking at it from academic viewpoint only. The system has yet not been made responsive to actual threats and vulnerability. There are no rules existing regarding the safety measures, procedures, infrastructure, practical standards in skills and the quality of response to situations. Reportedly, in 2005-06, there were 11 incidents of false threats of bomb implants in various educational institutions in Pune city alone. In all these incidents, the institutions could only take two actions - informing the police and leaving the students to go home. There was no effort towards safeguarding. No one can blame the institutions because no effort has been made to upgrade their safety, security and response mechanisms. Keeping a few security men (invariably ill-trained and unaware about disaster responses) is a false sense of security that many institutions and the corporate possess.

Threat Analysis

7.2 There is a dire need for every institution to conduct threat, vulnerability and risk analysis of own premises, surroundings, places where (and whenever) any curricular, co-curricular and extra curricular activity is organised. After such an analysis, preparedness and response mechanisms have to be instituted. A school gathering, a sports meet outside the school, a picnic or an educational tour and even school bus operations have to be analyzed in full and preparations carried out and procedures laid down and practiced for each type of disaster - natural or man-made, occuring during such operational activities.

7.3 **Disaster To be Catered for :** Disasters are categorized based on their intensities from L0 and L1 and one has to be prepared for all these.

7.3.1 Earthquakes.

7.3.2 Floods.

7.3.3 Tsunami.

7.3.4 Landslides.

7.3.5 Fires.

7.3.6 Road and rail Accidents while travelling.

7.3.7 Drowning.

7.3.8 Snakebites.

7.3.9 Bomb Threats.

7.3.10 Riots and other disturbances.

7.3.11 Building collapses.

7.3.12 Kidnapping.

7.3.13 Industrial accidents in the viccinity

7.4 Factors To Be Considered. The following factors have to be considered during the analysis:-

(a) Location of the premises and existence of degree, probability and intensity of natural and man-made disasters based on that location would form an important factor. Say, a prestigious public school that is located on a hill slope could be vulnerable to landslides. The strength of its building notwithstanding, the building may give up if the foundation collapses during a heavy landslide. On this score, the school authorities may not have any control, but safeguarding the inmates would still be their moral responsibility. Similarly, a school located closer to an industrial area is thretened of industrial accidents (say chemical fires and gas leakages).

(b) structural strength of the buildings to withstand quakes and floods.

(c) Structural and architectural design of the building to allow speedy evacuation.

(d) Material support available on the spot to mitigate any disaster.

(e) Number of trained personnel available to mitigate the disasters.

(f) Procedures that are laid down and rehearsals carried out to minimize casualties.

Mitigation and Management

7.5 The mitigation should be considered in three stages - Pre-disaster Phase (Preparatory Phase), 'During Disaster' Phase (Execution Phase) and Post Disaster Phase (Corrective and Feed Back Phase). Let us consider these in more details.

7.6 Pre-Disaster (Preparatory) Phase. This phase involves the following :-

7.6.1 Making Mitigation Teams. This is an important aspect of mitigation. The teams that could be formed depends upon the strength of the inmates in an institution, the type of disasters that may be encountered and the availability of the staff to form the teams. Normally, for any institution, a Rescue Team, a Relief Team and a Control Team should be formed. The possible structure is shown below: The teams that are formed are essentially from the teaching and non-teaching staff members. In colleges, NSS and NCC volunteers and some enthusiastic and physically fit students can also be co-opted. Where the structure of educational institution is complex, disaster marshals could be more in number - one for each floor or section and working under an overall direction of a Chief Disaster Marshall who should be the Principal or Director of the institution or may be a senior supervisor. The Control Centre Team should be made of office staff who invariably have communications and records available with them. The teams thus made should be announced, their duties laid down to cater to different circumstances and they should be collectively

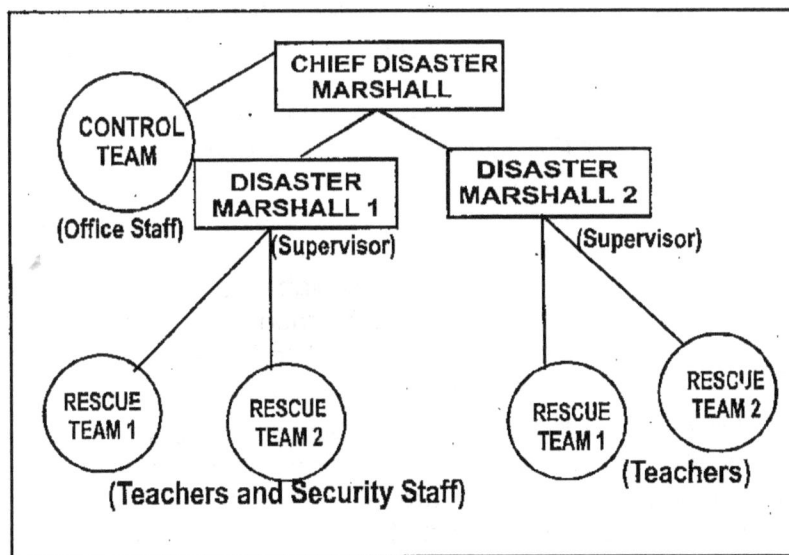

trained and practiced. Teams should be reviewed and restructured periodically.

7.6.2 Training the teams and the students. To executed any mitigation smoothly, training is highly essential. The training for different teams would have some common aspects of safety and security and specialized training concerning the respective teams as under:-

7.6.2.1 Rescue Training. This should include evacuation of the stranded victims, picking up injured from surfaces, emergent first aid, carriage of casualties and communications.

7.6.2.2 Relief. This training would involve first aid, administration of the victims (including resuscitation), food, water and lodging, psychological interventions and records.

7.6.2.3 Controlling. This would involve communication with government agencies, parents, recording of events, arrangements for administering the victims etc.)

7.6.3 Preparation of mitigation plans - 'on-site' and 'offsite'. Every institution has a particular location, structure and functioning. Thus, no two institutions could be identical. Each institution has to conduct vulnerability analysis periodically to ascertain the risks that the institution is likely to face during different situations. Once this is done, mitigation plans have to be worked out to face different calamities. The basic principles of each plan would remain same, however, execution would differ. Thus, the plans must include evacuation priorities and processes, identification of safe areas within and outside the premises, plans for dealing with fires, building damages, floods and even social turbulences. The areas that will be used and the facilities that could be made available have to be planned and placed. These plans should be then explained to all the teams in the form of Standard Operating Procedures (SOPs). These procedures must be reviewed periodically.

7.6.4 Precuring equipment. Mitigation could be efficient if the mitigators have certain equipment available to face situations till the situations are taken over by the government agencies. The level to which the equipment is required to be held depends upon the building structures and their complexity and spread. Some essential equipment includes a 100' long rope of 2" thickness, a 30' long ladder, a water pipe of 1" to 1.5" diameter, Foam and DCP type Fire extinguishers, Beaters and Bill Hooks, a generator set of minimum 2 KVA capacity and a water lifting pump of at least 1.5 Horsepower capacity. When the equipment is obtained and a minimum holding level is maintained, all teams should be trained in handling the equipment.

7.6.5 Keeping information at hand. For control Team to function efficiently, there should be certain static and some dynamic information system that should be operational. The static information must include telephone numbers of all mitigating agencies of the government, contact numbers of ambulances and nearby hospitals with their known capabilities, listing of various team members and their availability, Locations of the equipment and accessibility etc. The dynamic information should include the number of evacuees, teams operating and the progress of mitigation, number and possibly names of the trapped victims, disposal of the victims etc.

7.6.6 Off-site Responsibilities. When a school or college excursion is organized or when the school buses ply for collection and disbursal of students and the staff, an "Off-Site" intervention needs to be prefered. Such an intervention is also required when there are any programs organized by any educational institution away from its premises. The following aspects need to be borne in mind by the administration and the teachers and staff:-

❖ There should be a first aid box at in vehicles that are used. The drivers and the cleaners should be trained in offering first aid and must know actions during emergencies.

❖ The place of excursion must be analyzed for emergency evacuation and safety instructions should be imparted to all the participants. Some members of the staff should be adequately trained in emergency duties.

❖ Communication aspects during off-site activities need to be tied-up.

❖ Whenever there is any function at a location other than the institutional premises, the safety and security angles have to be analysed and catered for.

7.7 During Disaster (Execution Phase). When a disaster warning is received or when a disaster strikes without warning, the educational institute should quickly hail for the teams and the team members should firstly control the panic and immediately direct the victims to vacate unsafe places in orderly manner and move them to safety. The teams should quickly get together and work according to the SOP, fighting the disaster, evacuating victims from the surface, organize the relief measures; control team should inform the government agencies and call for medical help. The entire operations would depend upon alert and swift actions, based on the strength of the standard of training, practices and rehearsals carried out and education of the inmates of the institution.

7.8 Post Disaster Phase. This involves data collation and its analysis, upgrades in procedures to improve the disaster mitigation standards and review and briefings. This may lead to changes in the SOPs and equipment holding and may suggest better training of the inmates.

SAFETY AND SECURITY OF PLACES OF PUBLIC INTERESTS

7.9 Our country has a composite culture and a complex socio-politico-economic situation. We celebrate our festivals with such great ado, pomp and show that during the celebrations we hardly pay any attention to the disaster prevention and mitigation aspects. This has caused a great havoc in the past and our country is just about trying to grapple with disaster related social aspects. Be it Stampedes and Fires during religious festivals or be it the epidemics or terrorist activities, we have great threats to confront with and we need to take a realistic check on the way we celebrate our functions and administer our public places. Indian public follows orthodox rituals which may be outdated today. But a common man is not ready to change the ways of celebrations. In India, people may be more tolerant, but their religious sentiments get disturbed quite easily and hence the administration is reluctant to enforce stricter laws to avoid public wrath. This picture needs to be changed and the form in which the celebrations are held need to be changed.

7.10 Threat Assessment. The following threats exist during festivities and public functions:-

7.10.1 Threat of overcrowding in limited spaces, resulting in stampedes.

7.10.2 Threat of fire arising out of religious rituals or bursting of crackers or careless smoking and also unstructured electric connections resulting into sparking.

7.10.3 Spread of epidemics due to unhygienic conditions,

7.10.4 Terrorist attacks during the celebrations.

7.10.5 Collapse of temporary structures.

7.10.6 Vehicular accidents, cases of drowning or food poisoning.

7.11 Selection of Place of Celebrations. It is often found that many celebrations like the Ganesh or Durga festivals or some other birth centenaries are celebrated and religious functions and public rallies are held using roads. Temporary structures are aloowed to be erected, disturbing the traffic and creating congestions. Some decades back when the population had not exploded to the present extent and the traffic densities were lower, the occupation of roads was considered to be nothing out of the place. However, with the present situation, the criticality and threats are increasing to a great extent and we need to review our policy of erecting pandals for installing the deities and having programs on the roads. When celebrations are organized on the roads, traffic congestion and crowding of the public brings down the administrative mechanism. The blaring loudspeakers and bursting of crackers creates public nuisance apart from increasing the threats of fires and ability of the organizers to establish better control over crowds. To resolve this problem we need to allow such celebrations only in open spaces away from the roads-like the college or school grounds, bigger permanent structures of halls and may even think of shifting out. There are many temples located on hill-tops and their approaches are very narrow and crowd management is difficult. At such places, an alternative needs to be worked out by the society and religious heads should be made to understand the sensitivity of the problem. A time has come for us to redefine our methods to suit the present day society's safety and security needs. How should the place of celebrations or functions be? Firstly, it should be away from the thick of the traffic and not on any road. Secondly, the place should be spacious enough to accommodate the expected crowd at any one point of time and also considering the maximum rate of inflow of the devotes or visitors. The entries and exits should be specious and routes of evacuation should be created to ensure that during emergencies the rate of exit/ evacuation could be increased. Thirdly, the structures at the place of function should as far as possible be permanent and all efforts should be made to avoid erecting of temporary structures that could act as fuel for fire or restrict movement or collapse. Fourthly, the ingress and exit routes should be long enough to establish control. Fifthly, temporary commercial stalls should not be allowed near any place of function and if at all they are required to be erected, they should be at least 500 mtrs away from the main area of function and certainly outside the security perimeter.

7.12 Safety and Security Arrangements. In order that the place is secure and safe. Assessment of vulnerability and risks has to be made. Thereafter a plan should be made and drawn to the scale, to indicate all elements present at the site. The safety and security arrangements should involve the following:-

7.12.1 Arrangements for controlling the flow of the devotee with barriers and check posts correctly indicated. The plan should be approved by the civic administration before permission to hold the function is given and the arrangements should be inspected by the officials at least 24 hours prior to the commencement of the functions.

A Schemmatic Representation of A Religious Place

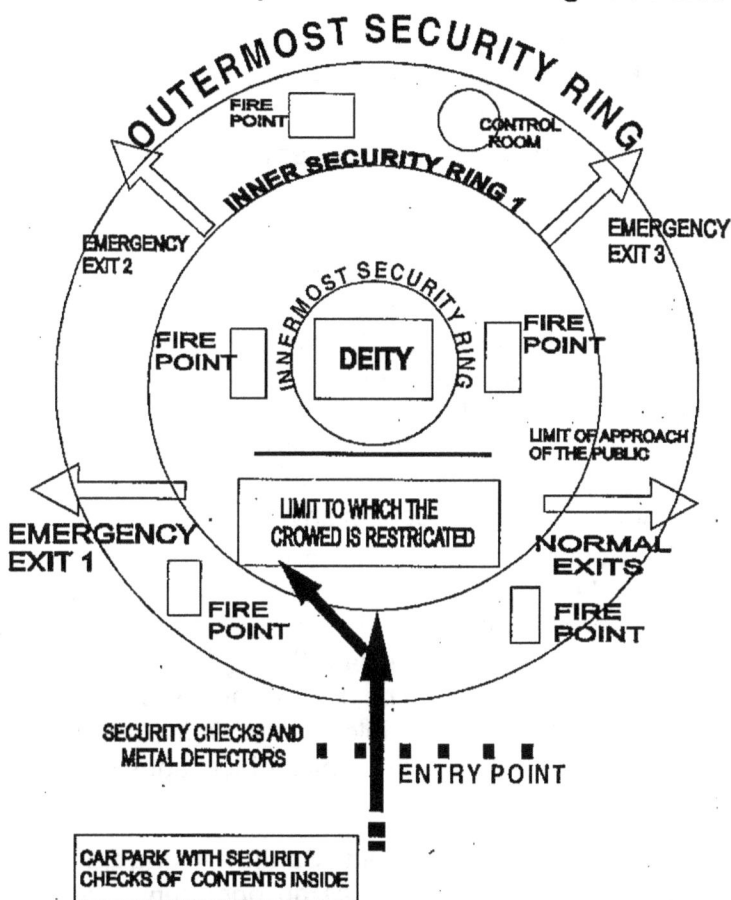

7.12.2 Arrangements for screening of people to avoid terrorist attacks should be a regular feature for all functions and at every place of worship. The places of worship should be accordingly graded.

7.12.3 Arrangements for fire fighting. The place must have a dedicated fire tender with sufficient water supply. There should be hand held fire extinguishers of the right type and to adequate quantity as well as fire beaters, bill hooks and water buckets. Such fire points should be suitable placed. There should be a trained pool of volunteers from amongst the organizers who should be available all through the celebration time.

7.12.4 There should be heavy first aid boxes with trained personnel to handle the casualties. Availability of a doctor during intense periods is highly advisable.

7.12.5 A public address system for crowd control is essential.

7.12.6 The place should have an inner and outer security perimeter with adequate security personnel trained in emergency duties and equipped with communication sets.

7.12.7 The religious rituals should be scrutinized for the way devotees offer certain articles to the deities. The place of offerings should be well outside the inner perimeter. Lighting of oil lamps without fire precautions should be curbed and the devotees should not be allowed to pour oil directly in the lamps or light lamps or break coconuts anywhere.

7.12.8 During mass flow at the time of celebrations like Kumbha Mela, there should be multiple points for bathing. Everyone should not throng at the same "Ghat". That has created problems in the recent past. During Ganesh emersions, every procession should not be directed at the same emersion spot. There should be multiple points, well controlled and multiple emersion points and no major blockages will thus be confronted.

7.13 Problems of Hygiene and Sanitation and Epidemics.
Celebration time is treated as an opportunity for casual hawkers, selling food items. Most problems occur due to contaminated water and food items prepared under unhygienic conditions. The administration is so much pre-occupied with the security and other arrangements that it hardly finds the resources to check the casual hawkers and the quality of water and food that is distributed. An administrative system should be created where proper permits are issued for food and water

distribution at select nodal points. No unauthorized hawkers should be allowed. In India, we have a problem here-first is the deprivation of small hawkers of their business opportunities and second is that of possible corruption. Our systems need to address these problems seriously. Preventive medication is just one of the many answers.

7.14   What is true for religious functions should also be true for exhibitions, major sports events and political rallies, in relation to safety and security arrangements.

7.15 Fire Crackers. Firecrackers have emerged as a big nuisance. Playing with crackers on the roads is rampant in our country. Bursting crackers above a particular strength of the explosion and pyrotechniques of low quality should really come under ban. There may be resistance from the cracker factories, however, in view of the threat that the users of fire crackers pose, the ban and control is worth. Every year, not only fire crackers factories report fire accidents, but during festivities lot of accidents occur. We need to realign our mode of celebrations.

7.16 Creating Teams and Layouts. For better control, the following teams and additional external resources should be deployed as shown in the diagrams below:-

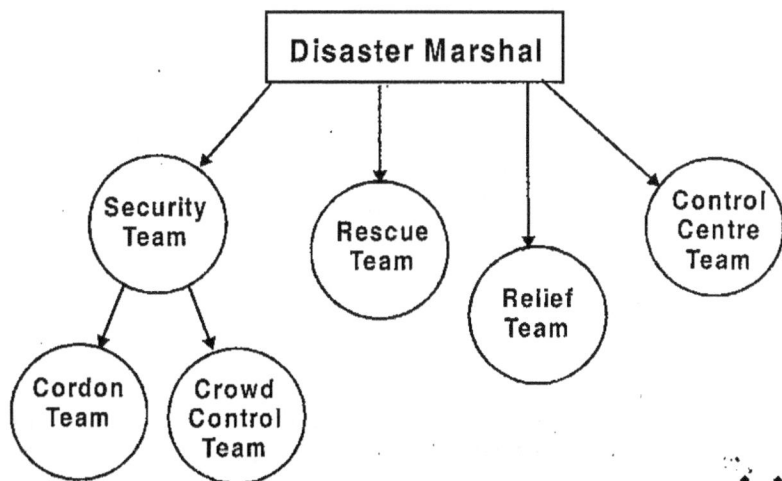

CHAPTER 8

LESSONS LEARNT FROM THE PAST

8.1 Every Disaster brings out certain aspects of successes and failures. The success manifests due to effcient systems, good awareness, presence of mind of the rescuers as well as victims and sheer acts of dare devilry.Absence of systems and awareness amongst the population or onset of panic and non-adherence to the rules apart from many other drawbacks lead to great destruction resulting in heavy liabilities.

8.2 Amongst the most severe disasters are earthquakes, floods, fires and road / rail accidnts.Here,we try to analyze some previous disasters that had taken place and try to see what went wrong and what was right.Through this analysis,we would be able to draw lessons.

Case1 :Tsunami of December 2004 :

8.3 On 26thDecember 2004,an earthquqke occurred in Indian Ocean off the coast of Indonesia,near Aceh at around 6.29 in the morning.The quake measured 9.1 on the Richer scale. Within minutes, a gaint wave was triggered . The wave measured almost 12 meters in height . The wave travelled at a speed of almost 800 km per hour and engulfed many coastal areas of Indonesia and Andaman and Nicobar group of islands within a few minutes. The Government of india got the information regarding this disaster immediately and the news was flashed across the country. However, by 8.30 a.m. coastal belt of Tamil Nadu in india was hit by the giant waves. The waves hit Kerala coast by 12.30 p.m. The waves spread inlands to almost 1 km destroying and damaging structures and washing off people. Thousands perished and many more became homeless. One peculiar effect prior to the arrival of the giant wave was that sea water suddenly receded by almost 500 to 1000 meters. Through people saw this happen, they never visualised that this recession of water was a precurosor to onset of a giant wave. The time gap between the recession of water and oncoming of the phenomenon, probably they could have gone to safety. This did not happen.

8.4 Could we have avoided this Destruction?

The earthquake and the resultant Tsunami could not have been stopped. However, loss of life could have been avoided by ensuring good warning systems. There are satellite and electronic warning systems existing in the world. The following lessons are learnt :-

● There is a need to deploy technological means for generating warning at Natinal and State levels.

● To ensure that the population gets warning in time, local warning systems need to be deployed, reharsed and kept operational at all times. These warning systems should be in the form of sirens and Public address systems established through police wireless network as well as village panchayats. It is not enough to give warning to the public. The public should also be instructed to take safety measures. If evacuation plans are in place, the warnings should be accompanied by executive instructions for immediate evacuation.

● Lack of knowledge by people resulted in people not going to safe places before the onset of waves.

● The development norms in our country in coastel areas are not in place and are not followed strictly. The construction closer to the coast must have norms such that each structure is able to withstand the most probable disasaters like tsunami, cyclones and floods experienced in coastal belts.

Case 2 : Earthquake of Bhuj (Gujarat)

8.5 Gujrat Earthquake of 26th January 2001 brought out many aspects of how human interventions increased the intensity of damage during natural calamities. No one could have stopped the earthquake from occuring. But, had we kept in mind the basic parameters of development, the destruction could have been much lesser. Some aspects are given in the subsequent paragraphs.

8.6 III- Convinced Develpment. The following aspects stand out :-

- The streets and lances were very narrow and buildings that collapsed took heavy toll of school children who were on their 'Republic day' celebration march.

- The construction norms should have catered to "Threat Analysis", as the affected area was vulnerable to earthquakes and structural norms for erecting quake resistant construction should have been followed.

- The builders had also used sub-standard material and ill-conceived designs during construction, thus causing greater damage to buildings.

- The destruction was so great and at so many places that the government agencies were grossly short in dealing with rescue and relief needs in the first 24 hours.

- The disaster planning and coordination was wanting. The government agencies, medical units and relief agencies within the area and within the state were highly inadequate. The lack of coordination caused administrative hassless during the first 72 hours and it was only thereafter that the government and outside agencies started taking control of the situation.

8.7 How Could We Have Minimised the Loss of Life?

- By incrasing the awareness level of people about actions to be taken during any disaster and forming teams of people to executive immediate rescue. Through, people did attempt to rescue each other, due to lack of training and means, their rescue efforts were not effective to the required extent.

- Any mitigation can be succesful if prior coordination exists within the governments agencies and the NGOs and proper resource planning and placement is done during pre-disaster phase. Had this happened, the losses could have been minimal.

- Municipal Corporaticns, Municipalties and the village panchayat could have ensured stricter development norms. This would have avoided collapsing of many buildings.

Case 3 : Flooding Mumbai in July 2005

8.8 Mumbai experienced unprecedented rains on 26th July 2005, with 900 mm of rain pouring down in a limited area within 24 hours. The entire town was paralyzed, many lost their lives, flood waters washed away lot of property and many industries suffered from loss of material. Secondary effects showed up within a few days in terms of a major landslide, some building collapsed due to rain and spread of Leptospirosis. Overall the loss was estimated in crores of rupees.

8.9 Salient points. The damage was caused due to the following factors :-

● Mumbai's drainage system was not geared up to the amount of water that was accumulated due to rains. The drainage chocking had occured due to accumulated waste in the drains and 'Mithi River'. Also, high tides brought reverse water flow the sea into the drainage system thus causing further accumulation of water on the roads.

● Waste and filth that had blocked the drains had started running down the streets and lanes spreading dreaded

● Illegal construction of hutments on the slopes of a hill at " No 3 Khadi" caused loss of life as the huts were destroyed in a landslide. A total of 140 huts were destroyed killing about 90 people.

● Building that had collapsed were mostly structurally weak and the tenants had refused to vacate them inspite of warnings given. In some cases, no efforts had been made towards inspection and repairs of the buildings.

● There is a need to estimate threshold levels of rain after which flooding starts in any area. A warning 'threshold' also needs to be set.

Case 4 : Fire in a School building at Kumbhakonam

8.10 In Tamil Nadu, at a place called Kumbhakonam, a fire broke out in a shool. The cause of the fire was treaced to the canteen that was run in the school building. The fire started while the canteen staff was cooking. The fire soon started spreading. The teachers panicked and instead of moving the children to safety, a stampede started. This fire resulted in the death of about 38 children and some teachers.

8.11 Points to ponder over :-

- The structure of the school building had restricted exits.
- Fear gripped the teacers and the students. No one could use common sense. The teachers did not take control of the situation and could not carry out an orderly evacuation.
- Access from outside was very limited and nobody could think of creating an access by breaking windows and effecting evacuation. Wooden staris had burnt down thus further aggravating the problem.
- The school building was not constructed keeping in mind safety and security standards.
- Any building must have emergency exists. Such exits were not available.
- The teachers were not trained to handle such disasters.

Case 5 : Fire in a School Function

8.12 At Mandi Dabwali, in Harayana State of india, a school function was organised Since the school did not have a covered hall, the function was organised on a ground, under tents ("Shamiyana"). This Shamiyana had only one entrance. The electrical connections that were temporary and the power loads were not calculated properly. These was a huge crowed of students, teachers and parents. During the function, due to electrical sparking, the tent cloth caught fire. The crowed panicked and everyone rushed towards the only existing exit. The chairs obstruvcted movement. The burning shamiyana spewed smoke as some plastic chairs started burning . In this fire about 450 people died, mostly students. The deaths were more due to asphyxiation than due to burns.

8.13 Points to Note :

- Temporaray structure always have a greater threat of fires.

- When electric connections are mede, proper socket- plug arrangements must be done and loads must be correctly calculated.
- In a cloth structure, one must cut the side walls quickly and allow smoke to go out. This also creates an additional exits.
- Stampede should be properly controlled. A rush creates more blockades.
- Public awareness should be improved.
- Smoke is as dangerous as the fire. Use of handkerchief, especially a wet one should be made to tie-up around the nose. This prevents suffocation at least for about 10 minutes.
- Fire fighting equipment should always be catered for whenever there is a huge crowed likely to assemble at one place.

Case 6 : Fire and Stampede at Mandhardevi Temple

8.14 In 2005, an yearly religious festival was being held at "Kalubai" temple at Mandhardevi, situated on a hill range in Satara district of Maharashtra. On the day of the festival, the crowed was more than what could be normally accomandated. There was only one narrow access route and it was further narrowed down due to temporary shops that sold material for offerings to the deity. The temple had a pillar with multipal oil lamps and the worshipers normally pour oil in the lamps. Coconuts are broken and offered to the deity as a ritual. On that fateful day, the coconut water had made the floor wet and became slippery due to spilled oil that go mixed with the coconut water. On 25th January 2005, the final day of celebrations, the diety was to be taken in a procession in a small wooden casade. A woman who as carrying the casade slipped over the floor and that triggered stampede as the devotees attempted to pick up the casade and bent down. The crowed was unmanageble and tha limited space made the devotees lean onto the shops. In this whole melee, the oil caught fire and the devotees started running helter-skelter. The temporary shops also caught fire. The locals who organised the festival had no means to control the mob. The devotees who had come from far distance started clashing with the administrative staff. Within minutes, the place turned into a violent frenzy with fire raging. There were no fire fighting arrangements and this caused hundreads of deaths.

8.15 Points to ponder.

- During any religious festival, when the crowed is huge and the space limited, crowed movement has to be controlled with strictness.

- Awareness of the population to differentiate between Religious celebrations and blind faith rituals needs to be improved.

- Temporary structures should strictly be banned. If at all this is not possible, proper fire precautions have to be instituted.

- Medical facilities, Fire Bridge and presence of police Force goes a long way in preventing mishaps.

- Access routes have to be critically examined.

- Administrators and trustees of various shrines have to be made aware and organisations at the site has to be planned and executed.

Some other Disasters

8.16 Burried under an Avalanche (Excerpts from the Reader's Digest September 1990). On 25th December 1988, Brett Woods and Keith Catcher decided to climb a steep 150 meters hill at the town of Durango in Colorado, USA. This was a lonely holiday for the two friends. It had snowed the previous night and 45 centimeters of fresh snow had piled up on the hill-slope. the hill slope had a road next to it, going through to the town. They travelled in Brett's car to the hillside and after parking the car, got down at 3.30 pm. They had planned to climb 30 meters when they were out of breath. They sat at a spot to get their breath back. As they sat, thay heared a sharp cracking sound and looked up. An avalanche had started coming down the slope towards them. Before they could rise up to go to safety, the falling debris of snow engulfed them. They were buried, not too far from each other. Coughing and struggling to get fresh air, fear gripped them and the fear turned into panic. Frantically, Brett tried to move his head back and forth to create a tiny hole of air. He tried to move his hand, but coulden't. The weight of the snow had trapped him underneath. Brett shouted for Keith and after a few shouts, got a feeble answer. From the sound, he could guess that they were not very far from one another. They both tried to move closer and succesed in touching each other's boots. That was a psychological support. Realising that they would suffocate and their only chance lay in freeing themselves. Deseperately, They tried to dig had started bringing down their body tempreture and they both had to rest very frequentely. They kept shouting each other's names to ensure that psycologically they were not getting drained out. After some time, Brett managed to clear a hole through which he could see the sky. But, the snow was getting hardened and it was not possible to dig a bigger hole to free himself. Brett was shouting for Keith, but as the time passed the response from Keith had started becoming irregular. Brett could hear cars moving on the road that was quite near, but his shouts were not heard by the passerby. It got dark and cold. Brett feared that they both coulden't sustained through the entire night. Brett shoved his bright blue snow glove outside the hole and waited for some sounds of the passerby. He heard a couple walking the road and started shouting. The couple heard the bleak sound of "help" and started looking towards the slope. With a torch, they started climbing and searching for any tell-tale signs of somebody in distress. They spotted the glove and reached there. Hearing their talk, Brett again shouted. The couple located Brett and assured him that he will be dug out. But, the hardend snow was difficult to clear. The couple went to the road and waved the passing vehicles to stop. With the snow shovels, they started digging. A car driever radioed the town for police and ambulance. At last Brett was rescused. He immediately pointed out to the hospital and recovered in due course of time. It was a miracle that they both survived low temperatures and the trauma.

What really saved them was their psychological support to each other, a strong sense to live and their efforts to free themselves and create maximum space for breathing while the snow was still soft. Brett and Keith came out of the disaster as winners.

8.17 Overboard in a White Sea. (Excerpts from the Reader's Digest, July 2005).

A german container ship Hansa Bergan was sailing from Singapore and bound for Maturitius, was lurching through heavy storm in the Indian Ocean. The waves were eight meters high and spewed spumpes of water. Second Officer, Kerstin Burns, the only woman officer on board was ordered by the Captain to help secure the gangway that was breaking loose. It was 3pm on june 25th 2004. On the deck where Burns was trying to help secure the gangway, a giant wave rose up and splashed over the deck. Before burns could secure herself, she was swept off her feat into the raging sea. By the time the water subsided from the deck, the other crew realised that Burns was missing. The ship was sailing at 25 knots per hour and it was difficult to even locate the young officer. "person Overboard", went the alaram and all hands were ordered to the deck. The ship was ordered to turn. The Captain, Helmut Wende, noted the location of Burn's washing off and radioded to them. In the sea, Burns tried to swim towards the ship. She was only praying that no sharks should be heading towards her. The ship that had turned back had started searching the area in a square formation. The waves made Burns ride a crest one moment and go into a through the next moment. Whenever the ship approached close by, she tried waving at it with the hope that some one would notice her. But, it was difficult to locate a tiny spec of a human amidst the raging sea with wind velocity over 50 knots. After a few hours, the other two ships joined Hansa Bergan in the search. They dropped smoke flares and then light flares as it grew dark. Burns did not give up. Her body had started aching. yet she did not give up. The search continued the whole night. The ships were not ready to give up the search so soon. At the day break, Burns was still swimming and the ships were still searching. Burns saw one of the ships heading directly at her and was scared that the ship's water discharge would suck her underneath. She frantically tried to go aside. A wave made her ride the crest and she saw the ship just a few meters away. she waved. It was at this time that the alert crew noticed a black spot in water, very close to her. careful observation confirmed that the spot was only a human and the human was attempting to swim and remain afloat. Overjoyed, the ship radioed the message to others. Now efforts began to pull her out of the sea. It was difficult to do it- if the ship went too close, there was a danger that Burns would be sucked under the ship's hull. The distance to throw rope ladder had to be right. After many attempts, finally, they succeeded in pulling Burns abroad. Burns had been in water for nearly 19 hours, fighting against the giant waves. She spent only one day in a hospital and was sent on leave. She soon resumed her duty, more determined.

It was determination of Burns, never-say-die attitude of the searchers and a great faith in the heart that Burns survived. Indeed, Human courage knows no bounds.

8.18 Trapped in a Forest Fire (Excerpts from the Reader's Digest August 1990).

It was saturday, August 20, 1988. Pat hedges 35, along with his children, Kathleen 16 and Keith-15, and their family dog Misty was on an outing in a forest bordering Yellowstone National Park on the slopes of Horseshoe Canyon in Montana. After spending an adventurous evening, they pitched up their tents a half kilometer away from the Horseshoe Creek, amidst the pine trees. They could see an amber light, about 16 kilometers away, probably of a forest fire that was burning for a few days. "It is quite far off and will not travel up to us" thought Pat. The family had just been asleep for an hour when some cracking sound brought Kathleen awake. She sat up when she saw flames, a few hundread meters away. She called for her father, a captain in the fire department. When pat saw the flames, he was surpriced to see that the forest fire that was 16 kilometers away had crept so fast to the section where they had camped. Pat could see smoke rising up and soon felt the heat. He knew that they had to react fast and the speed of burning pines would soon overtake their opwn speed of running. Pines offer a good natural fuel to any fire. Pat had to decide fast. He ordered his children to dash down towards the Creek half a kilometer down slope. And thay all had to beat the oncoming fire because the fire was directly crossing their way. There was no time to think anything else. The family dashed down the slope and were just able to stumble their way upto the creek when the entire area around them was in flames with pines falling down with exploding sounds. Amidst the falling pine needles, they managed to dive into the trickle of water and submerge themselves. The water was freezing cold and yet they coulden't dare to peep out because the raging fire was threatening to burn them. They had carried their sleeing bags to protect themselves. The smoke was a ghastly experience. However, they kept their cool and were determined to beat the fire. The ground oputside the small creek was also hot with burning pines and the undergrowth. It was only a few hours before the fire burnt down everything in the area and the family could venture out. They found their way back through an arduous trek of almost 15 kilometers, to the safety of their home. It was quick thinking and speedy action with good presence of mind that saved their lives. They remembered that water and cooling effect was their only savior.

8.19 Gas Cylinder Explosion.

Mr. pendharkar, an engineer, was a partner in fabricating business. His company was given a contrast by a sugar factory at shrirampur in Maharashtra, to construct three employed a young graduate engineer and himself used to visit the site twice a week for supervision and guidance. On 24th February 1984, Mr. Pendharkar was on his routine site visit. He found that the cladding was not approprite and he adviced the young engineer, Mr. Gaikwad, to remove the incorrect part of the structure and re-weld it. Pendharkar himself laid out the metal sections and explained to Gaikwad and the workers the actual position. The welders were working on acetylene gas cylinder for welding the metal segments. As Pendharkar and Gaikwad were discussing some matter, abuot six meters away, one of the cylinders exploded. Two workers were killed on the spot. Clothes of Pendharkar showed presence of mind and dragged Gaikwad to a nearby Water tank that had been constructed for storing water for the construction. Pendharkar pushed Gaikwad inside the water and himself jumped inside the tank too. The fire was extingunished immediately. However, the burning sensation persisted. Skin from their bodies had peeled off at many places. Pendharkar's clothes were embedded in the flesh as he was wearing synthetic clothes. One of his venis near the neck was also cut and there was a profuse bleeding. One of the workers managed to hail a car and Pendharkar, along with Gaikwad were taken to a hospital at Miraj. Timely treatment saved their lives. They both had suffered from more than 60% burns. However, they were restored to normalcy with the efforts of the doctors.

The analysis of the above incident brought out many factes. Firstly, handling of the gas cylinder is a major factor that needs to be kept in mind while operating with such equipment. Secondly, wearing of cotton clothes instead of synthetic clothes is very essential when working on sites. Head protection is important and availability of fire extingunishers and water at the site is inescapable. Thirdly, presence of mind saves lives. Pendharkar did not panic even when his own clothes were burning. Lastly, immediate rescue and removal to a competent medical authority is important.

8.20 He Kissed the Death. In November 1986, at Binaguri Army Cantonment of West Bengal, Major Marathe (The author of this book) was away on a field excercise with his brigade. His wife, children and mother had stayed back as the schools were still in session. At 8.30 a.m. that fateful day, his two daughters had gone to school. His wife, mother and a seven month old son, kedar, were at home. While his mother was having of a family welfare function and young Kedar, who had learnt to crawl was playing around with his toys. After some time, Mrs. Marathe noticed that the young lad was missing from the room. Worried, she excused herself from the telephone talk and searched around the house. The entrance was locked and thus, there was no chance of the kid going out. She couldn't find Kedar anywhere in the five room flat. She peeped into one of the bathrooms and was shocked to see Kedar toppled in a bucket half filled with water. The boy's head had completely immersed in water and there was a feeble movement of the legs. She quickly pulled the boy out of water and found that his eyes were rolled up and his breathing had stopped. Panic gripped her. But, she put her young son upside down on the carpeted floor and started pumping his back. she also shouted for help and her mother-in-law quickly came to her help. Hey both held the boy's feet and holding him upside down, started gyrating him around. Her shouts brought the neighbours and she requested them to call for an ambulance. She and her mother-in-law then ran down on the road to fail for any passing his back. Every second was also trained in first aid. Though fear had gripped them, they did not leave their balance. soon, young Kedar started coughing and showed irregular but shallow breathing. They continued to rub his back. Lot of water had gurgled out of his mouth and the nostrils. The two ladies managed to stop a jeep and took kedar to the emergency ward of the military hospital. By the time kedar had reached hospital, he had started breathing, his heartbeats were normal and he was only crying feebly out of fear. But, he was totally out of danger. After spending a day under observation, he was allowed to go home with his mother.

It was a touch and go affair. The boy had gone very close to death. Another 30 seconds of delay would have destroyed a young life. The credit for saving her own son was entirely that of Mrs. Marathe, duly assisted by her mother-in-law. Presence of mind, prompt action, controlling own fear and thinking of right actions under stress were the reasons of Kedar's survival. He owes his life to his mother in the true sense.

8.21 Thrown in a Ravine. On 1st March 1997, Mr. Narayan Alavani prepared to go to Mumbai with his family. Mr. Alavani resides at Pune. Himself a car and scooter mechanic, he had a scooter with a side car. A scooter with a side car is not the right transportation in such a journey. However, he was confident that if he drove slow, he would make it to Mumbai without any problems. At 11 a.m. the family was in the hill section at Lonavala when the outer wheel of his side-car got detached and Mr. Alavani lost his control. The scooter went in a deep ravine. Vinay, Mr. Alavani's son and Varad, the grandson, were thrown on a flat piece of ground. Vinay was bruised and shocked, but when he saw Varad, crying, he realised what had happened. He looked around for his parents. He found his mother unhurt, but in a deep shock whereas his father was bleeding, incoherent and disheveled. Vinay tried to pick up his father, but he couldn't. Vinay then pacified his mother and climbed up to the highway to seek help. No driver was ready to halt. Finally, Vinay decided to lay down in the centre of the road. That proved effective. A jeep driver stopped and on requesting, offered help. The jeep driver and Vinay could not move Mr. Alavani to the road side. Finally, the jeep driver dropped Vinay at Khopoli and there, two people helped him. They travelled to the accident spot, thereafter called the highway police and the ambulance and managed to evacuate everyone. Mr. Alavani had a fracture in the skull and had to be operated. Mrs. Alavani was evacuated to Sion Hospital at Mumbai and then transferred to pune where he was successfully operated upon.

One must understand that while planning any long drive, the mode of transport has to be right. Vinay showed courage and presence of mind when he forced the jeep driver to help him. The highway police should have been approached earlier.

Conclusions

8.22 In any disaster, however big or small it be, control over panic, presence of mind, a prior Knowledge of the rescue and first aid does help a lot. While planning any major event or even through routines of the day, safety and security aspects should always be kept in mind and catered for. Disasters don't always give warnings and one must always be prepared to face the unexpected.

❖ ❖

CHAPTER 9

PRACTICAL TIPS IN DISASTER MITIGATION

9.1 Disaster Mitigation is something that covers a wide arena of elements - permeating from International Level to a common man. Here, we deal with tips that could be useful to a common man during day-to-day affairs.

First Aid

9.2 First Aid Boxes should be held in every home, institution, industry, transport modes, public places and every other place where any organisation exists - formal or informal. Every human above the age of 12 should be made to undergo regular training and it should be introduced as part of the curriculum in every educational institution, as part of drill in every public/private sector or NGO. The training should be graded for different categories.

9.3 Contents of First Aid Box. A First Aid box for medicines of daily use and the one for mitigation a disaster should be separated. A First Aid Box aimed to mitigate a disaster should have the following contents :-

- Four Triangular bandages - 1 mtr wide at the base, made of cotton fabric.
- Two 2" wide rolled and autoclaved bandages.
- Cotton bale.
- Scissors
- Elastoplasts tape.
- Cotton Gauge
- Antiseptic Lotion for cleaning wounds
- Distilled water bottle
- A one-mtr long lace for tying a tourniquet.

- A pencil of 6" length
- A permanent marker pen to scribble the patient's administered treatment for a doctor to know. This has to be marked on the patient's forehead.
- A torch with cells.
- Two sticks, 1" thick and one foot long (for making a splint in emergency).

Information System and Organizing Home/Office/ Organizations for Emergencies.

9.4 Home. Each home must have the following emergency information system and markings:-
- Telephone numbers of all members of the family - office, cell phones.
- Telephone numbers of the immediate neighbours and most close relatives.
- Telephone numbers of Nearest Police Station, Fire Brigade, Heart Brigade (If existing in the town).

9.5 Each member must inform his or her likely schedules and whereabouts of the day before leaving home. This will facilitate the other members to keep a tap in case of emergencies. Each member must told the information of nearest safe plae- a relative's or a friend's residence or office in different areas of the town. During any emergencies, they must approach those friends and relatives.

9.6 Every office/institution must hold information about all the employees which should include the telephone numbers of their next of keen and any other close relatives, about their blood groups and somebody from the office should have seen the residence of each other employees. This should be preferably done department-wise.

9.7 Other Interventions.
- In the offices and homes, the fire extinguishers and first aid boxes should be prominently displayed and

inspected regularly for contents and their validity.

- The most important items that have to be saved must be kept in a separate cupboard such that the entire cupboard could be removed early or the contents could be emptied out immediately. All members of the family should know the locations of such things. These items should not be heavy and should prefearbly be indentity documents, passports, bank chequebooks, some cash and any other important item of lesser weight. Documents concerning ownership of property, any other business deals and fixed deposit receipts etc could form part of this bunch of important items.
- One must ensure that 'trash' is regularly disposed off and the premises should have the least combustible material.
- Electric connections should be regularly checked for sparking and leakages. Gas Regulator knobs should be switched off when not in use.
- A bucket of water should always be kept filled and handy in the house and in offices.
- Balcony grills must have emergency escape windows that are lockable. The keys should be kept handy and secured. In the offices, emergency exits should not be cluttered with other items and their lock keys should be kept at hand.
- Each office, institution and housing complexes must preferably have emergency alarm systems fixed for warning all the occupants. The alarm switches should be providded parallel in each section or house, to be operated manually. This should be similar to the one fixed in the banks.

Actions In Emergency

9.8 In case of Fire.

- Find out location of fire.
- Shout "FIRE-FIRE" loudly and call for help.
- Check the people endangered and remove them to safety.

In case own self or any other person in engulfed in fire, wrap own self and the other person with a thick blanket, bed covers, table covers, curtains etc to extinguish the fire on the person's (or own) clothes and immediately pour water on the person or self to bring down the temperature. Do not remove burnt clothes from the body - it may tear off the skin and flesh also and cause more damage.

- Use fire extinguishers if available. If fire extinguishers are not available, use material like corn flour, atta etc to suffocate the fire. Use of such material is suitable when the fire is small -like initial sages of cooking oil catching fire.
- If the fire is beyond local extinguishing, simultaneously ask someone (or do it himself) to call for the fire brigade and ring emergency alarm.
- Remove combustible material from around the fire. Also remove important items from the cupboards if marked as such with the help of others. However, to do that, one should not risk life.
- Wet the walls and the furniture with water with the help of other helpers.
- At no stage should one enter the fire alone, without warning others and without protecting own self with blanket etc and wetting own clothes before entering the flames and smoke.

One must tie a wel handkerchief over own nose to protect from suffocation. This may be a temporary measure in emergency for 10 minutes or so.

9.9 In Case of Earthquakes

- If on the ground floor, run out into the open, at least 20 mtrs outside the building.
- If on higher floors, try to stand under a strong beam and column joint or door arch, preferably on the outer flank of the structure.
- Protect your head with hands.
- Grab a table and create a slanting protection overhead and sit under it.

- When debris is still falling, don't venture out of the protected place. Earthquakes have after shocks and lot debris falls during that time.
- Shout for help and indicate your location. Also, attract other victims and appeal them to collect together at a safe place.
- When the debris stops falling - may be after a few minutes, venture out of the protected place, into next better protected place. Cover your head during the move and move slowly. Don't make a sudden rush.
- Move towards outer flank and try to descend down to lower landings.
- Help trapped people inside the building.
- If you are a volunteer and want to help in rescue efforts, make sure that you have adequate training and conduct the operations under supervision of a competent authority.
- Try to switch off gas knobs of the regulators if you are in kitchen. The earthquake may result into secondary effects resulting into gas leakages and subsequent fires.
- Do not try to fight a fire in debris.

9.10 In Case of Floods.

Except in case of the flash floods, warning is generally available. The degree of precipitation is indicative of likely floods. Where population is situated at the base of hilly regions, they would have to base their judgement on the basis of rains in the immediate and catchment areas. Today, flood control measures and constant weather monitoring and monitoring of water levels in rivers has made predictions to be possible. Considering that a warning of about 24 hours is available to the population in the flood threatened areas, the government interventions would be adequately built up in terms of evacuating threatened villages and localities and sheltering them temporarily. The following points need to be considered by the threatened population prior to being shited to safer places:-

- Before shifting out, collect own important documents - passport, ration cards, gas connection cards, domicile and other educational certificates, documents concering properties, identification documents, credit cards etc and cash, cheque books.

- Make sure that the gas knobs are switched off, gas cylinders are disconnected and secured in a manner that these do not float etc.
- Remove all valuables to safer places, duly locked.
- All material that is likely to get spoiled should be packed in polythene sheets. This material should be correctly secured in cupboards.
- Take essential clothing and leave the rest packed in thick polythene sheets.
- Take along essential medicines of the patients under treatment and their medical case papers.
- Carry some hard variety of rations and safe drinking water bottles to last at least 24 hours.
- Use own transportation means under proper guidance of the government authorities. Shift other motorized transport to safer places and keep them locked at the houses of relatives in safer areas.
- When adequate warning is not available and you are caught in the rising floodwaters, move in the direction of higher-level grounds.
- Keep empty water bottles and cans at hand, to be used as floats. Individual emergency life jackets can be made out of six empty water bottles. Wooden cots can be very handy as rafts, if attached with some additional floating material like thick thermocol pieces, tubes etc. Keep a rope and stick at hand.
- Inform your whereabouts to nearest relatives and seek their help.
- The floods cause water borne diseases. To avoid being a victim, do take inoculations immediately after the floods. Special care should be taken of the old and sick people and young babies.

9.11 During Cyclones

On the eastern coastal belt of India, cyclones are a common phenomenon each year. Cyclones also threaten the west coast. Generally, a coastal length of about 500 km is affected with a depth of about 50 km experiencing devastation. Weather forecasting has made early warnings possible. Generally a warning of three days is possible with general direction of approach of any cyclone known. The cyclonic winds do change directions within those three days and exact extent of coverage is possible to predict at least 24 hours earlier. With this facility being available, the general population in the areas that ared likely to be affected to bear the following points in mind and act accordingly.

- If required, the public must vacate places immediately in the vicinity of the coast, as sea waves also serge in.

- Remain confined inside their houses. If the houses have weak structures and roofs, shift to stronger structures.

- Secure roofs and loose items of the structures.

- Close all windows and fix sticking tapes for strenghthening, across the glass panes.

- Ensure cooking gas and rations to last at least 48 hours are available in the house. Fill up enough drinking water.

- Keep emergency lights at hand and switch off hte main electric supply.

- Prepare food and keep it at hand to last for 24 hours. Switch off and disconnect the gas cylinders.

- Keep enough stock of medicines for the patients and milk for small babies.

- Keep first aid kits at hand.

- Keep cell phones fully charged and use them very sparingly. Keep battery operated radios for listening to news bulletins.

- Keep cars and other vehicles secured on the reverse

side of the building structure with direction of the cyclone kept in view.

- If feasible, cut the tree branches in immediate vicinity of the structures. Do not venture into the open within 6 hours of the cyclone's likely strike time.

- Be prepared to accept additional people who may seek shelter in your house.

- Remove TV aerials and dish antenas.

- The government must announce likely time of the edge of the cyclone hitting a particular area and the possible time of moving of the cyclone to the next area. Police wireless network and mobile public address systems should be used.

9.12 During Accidents

Accidents are a very common occurence. Almost one lac people loose their lives in India due to road accidents. Safety precautions while driving - use of seat belts and helmets, adhering to traffic rules and staying within the speed limits can not really be replaced by any other intervention. Apart from this, when accidents occur, timely anticipatory actions by the occupants of a motor vehicle, prompt reactions by the victims and tthe people around the site help in reducing the number and extent of casualities. Following are the guidelines during such cases:-

9.12.1 When Travelling in a Transport and Accident is Imminent.   •

If driving, and a head-on collision is expected, swerve the vehicle to the gutter side and apply breaks fully. Simultaneously, shout for other occupants to take safe crouching position. The driver also has to shift sideways towards the door to get out of the way of the steering wheel's impact.

If a co-passenger, take a crouching position to protect all the soft organs and head. Here, when strapped, it is not possible to bend down. It is better to bend the legs at the knee and lift the legs so that the knee level is in the line with the waist level or above. If cushions are being used in the cars, the same could also be used to protect own self. One must remember that the reaction time is always very less and a prompt action is expected.

In a public transport, there are incidences when the vehicle has a failure of brakes. If in a hilly section and going down slope, the driver needs to halt the vehicle by scraping the vehicle along the hillsides and warning the passengers. This ensures minimum damage.

Inspection of luggage in public transport and identification of passengers and their luggage has to be taken up as a serious exercise in any public transport. The bomb blasts in buses and trains suggest us to take preventive measures.

Inform the police, ambulance and any emergency rescue and relief services. Each public transport must have numbers painted in bold behind the driver's cabin and at the rear of the vehicle body. Public transport must have First Aid boxes.

9.12.2 When at the site of an Accident

When at an accident site, quickly gauge the situations. See whether the victim vehicles have any fuel leakage. Check whether fire has started. This will suggest the most critical flanks where threats could exist.

Assess the number of casualties trapped inside the accident debris and the casualties clear off the debris. The casualties that are clear of the debris should be taken away to a safe distance and first aid should be applied immediately - **breathing, heart functioning and bleeding** should be addressed immediately in that order and efforts to revive and stabilize the casualties should be done.

If more help is available, simultaneously try to retrieve the trapped casualties. Care should be taken that while taking the victims out, the victims conditions do not deteriorate. The rescuers performing such acts must study the debris parts to which the victims' bodies are in contact and ensure that the separation of the victims is done quickly and carefully. After the separation, go through the first aid drills.

Inform police, highway patrols, ambulances and any other emergency rescue and relief services.

Evacuate the casualties by the fastest means. While loading and carrying the casualties, it should be ensured that the casualties' bodies are well supported and are not allowed to hang or drag. Giving support to a casualty's neck and waist is very essential.

Placing a heavy first aid box in every public transport should not only be part of the rules, but implementation and regular checking should be ensured.

First Aid is Everyday Emergencies

What is First Aid? It is the help given to casualty to stabilize and revive at the place of the incident, with whatever material that is available at hand, till a medical authority starts treating the casualty.

9.13 When Casualty is Located.

- Inspect the casualty immediately for injury marks. Carry the casualty to a nearby safe place with care and place the person on a firm platform or ground.
- Straighten out the body carefully and inspect the casualty from 'Head' to 'Tow'. Check the pulse at the wrist and at the neck. Check breathing. Check for bleeding - external and signs of internal bleeding. Thereafter check for fractures.
- Get someone else to fetch an ambulance while you attempt to revive the casualty.

If the Breathing is feeble or completely stopped.

- **How do you check it? -** Place you cheek near the casualty's nose and watch the movement of diaphragm. If the casualty is breathing, your cheek will get the feel of the expiration of air. Also, the stomach movement will be seen. You can also feel the breathing by placing your palms on the casualty's chest lightly. If there is no breathing, then take following actions:

- Loosen the clothing by removing belts and loosening shirt buttons.
- Clear the air passage by tilting the neck a little rearward. Inspect for blocked windpipe and clear any obstruction with index finger, very carefully.
- Follow methods of artificial respiration and if required give mouth to mouth respiration. These methods have to be learnt through a formal training and practice. The most commonly used methods of artificial respiration are :
- Schaefer's Method - Best suited for the casualties that are removed from water. Here, the effort is to remove water from the respiratory track..
- Holger Nielson's Method - Best suited for the casualties removed from smoke
- Silvester's Method - Used for the casualties under mental shock.
- Sea-saw Method - Best suited for small babies.
- Mouth to Mouth Respiration.

Tilting the Neck : Place your hand below the casualty's neck and holding the casualty's chin, tilt the head softly backwards. This results into the tongue getting pulled back and not obstructing the air passage.

Schaefer's Method of Artificial Respiration: This is best suited for drowing cases where the casulty's air passage is water clogged. Firstly, the water is removed by placing the casualty with face down and then lifting the casualty's waist such that the head remains lowered. This allows the water to flow out. Thereafter the Schaefer's method is used to restore the casualty's breathing. In this method, pressure is applied on casualty's waist. This stimulates the diaphragm movement and restores the breathing.

Holger Nelson's Method : In this method, the casualty is placed lying on the stomach and head rested on one side. The 'First Aider' sits at the head and applies pressure on the shoulder blades (this stimulates the lungs). Thereafter the casualty's elbows are gripped and raised above to allow chest expansion, while pulling the elbows towards the First Aider.

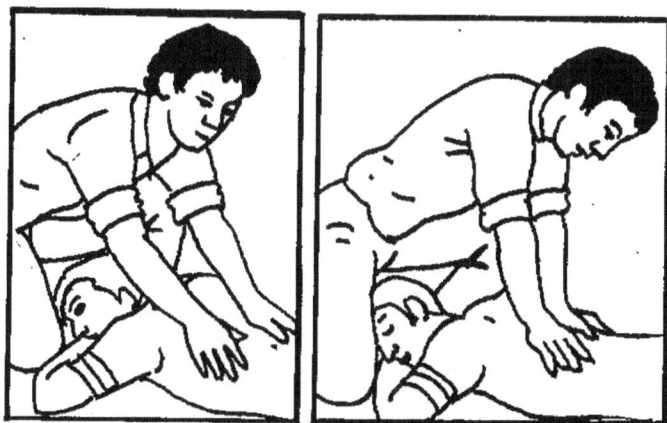

Silvester's Method: In this, the casualty is placed facing the sky. The casualty's arms are stretched on the sides. The First Aider holds the casualty's wrists and with a rotatory motion, brings both the wrists together and presses them on the chest and increasing the pressure counts upto three. Then stretches the hands to original position and repeats the process. This allows contraction and expansion of chest alternately.

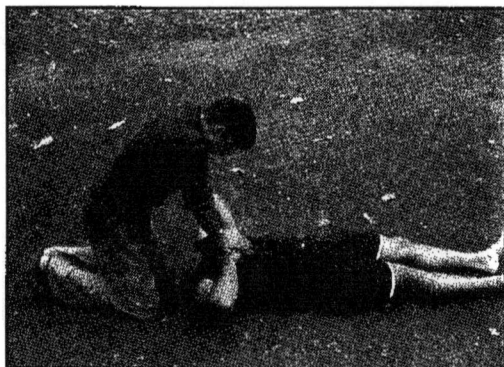

Mouth-to-Mouth Respiration : Hold the casualty's nosestrills in a pinch. Tilt the casualty's head backwards. Opening the casualty's mouth, cover it with own mouth and blow air inside the casualty's air passage. This process should be repeated at an interval of 3 to 4 seconds.

If the heart is not beating or the pulse is very feeble.

This indicates the need for giving a heart massage. Acquire a position with the casualty lying face up. The First Aider ssits with the knees resting on the ground on two sides of the casualty's waist. The pressure of both palms is applied and released in quick succession like a pumping action. For this method to be followed, the first aider has to be trained well.

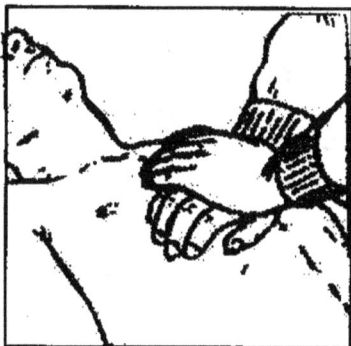

Dealing with Bleeding Wounds :

- **Severe Bleeding from the Limbs:** For this, the best option is to die a tourniquet on the upper side of the wound. If it is not possible to tie a tourniquet, raise the limb to restrict the blood flow. If there is a foreign body inside the wound, apply pressure on the edges of the wound and apply dressing on the edges of the wound. If the foreign body is small and not embedded in the flesh, remove it with forceps (if available) or with a cotton swab. When tourniquet is applied, make sure that it is loosened after every 20 minutes for a few seconds. Retie the tourniquet if the bleeding persists.

- On small wounds, apply pressure and tie bandage after the wound is cleaned.

- **Internal Bleeding.** It is very difficult to stop this and removing the patient to a hospital by the fastest means is the best answer.

- **Bleeding due to head injuries.** Keep the patient reclined with head up. Clean the wound and. Place cotton gauge and thereafter autoclaved cotton. Apply bandage lightly, covering the ears also.
- **Bleeding from Stomach or Chest.** Try to seal the wound with cotton and bandage.
- In all severe cases, make sure that the casualty is removed to the nearest hospital by the fastest means. Make sure that the bleeding part is kept raised up while placing the casualty in any vehicle.
- When blood oozes out continuously from the mouth, ensure that the mouth is cleaned frequently to avoid the patient from choking. Particularly when the casualty is unconscious, this care has to be taken.

Burns and Scalds.

- Do not try to remove the burnt clothes. Only extinguish the clothes.
- Immerse the burnt portion in water. You may pour water over the entire body so that the skin temperature comes cdown and scalding stops.
- Do not try to burst any blisters.
- Remove rings, bracelets, shoes and tight fitting clothes that are not burnt.
- Any burn larger than a postage stamp should be immediately referred to a doctor.
- Ice may be packed in a towel and applied to the burnt portion during the patient's journey to the hospital
- Cover up burnt tissue with light cotton gauge and cotton. Give fluids to the patient to drink while waiting for the doctor.

Carriage of Casualties

9.14 Carrying a casualty is as important as offering First Aid. Many times, due to wrong handling of casualties, the conditions of the casualties deteriorate. Also, recovering of the casualty from the place of incident and removal to a safe nearby place poses many practical problems. Some easy methods have been given in the subsequent sub-paragraphs.

Single Person casualty carriage methods: A casualty can be carried by a single person applying either 'Pick- a-Back' method, or a 'Fireman's Lift', a 'Drag' or 'Pull'. The methods depend upon the type of injury, the place of incident, the weight and height of the casualty and the strength of the person evacuating the casualty.

- **Pick-a-Back Method:** In this method, a casualty is lifted in a sitting position. The evacuator places both legs of the casualties on either side of own self and holds the knees of the casualty. The weight of the casualty is taken on own back. If the casualty is conscious, the casualty holds the evacuator around his/her shoulders for support. See the diagram.

Fireman's Lift: In this method, the rescuer lifts the casualty taking the casualty's entier body weight on of both his'/ her shoulders. The casualty's one wrist is gripped and the other

hand is either trapped inside the casualty's own hand or if the casualty is conscious, the casualty holds the lifter's shoulders or one arm. See the set of diagrams shown below.

● **'Drag' Method:** This method is used for bringing the casualty down the stairs when the rescuer is not able to use the above methods. See the diagram.

Pull Method : This method is employed when a casualty is in smoke filled room and the rescuer can not stand. The rescuer ties both the hands of the casualty and places own knees on both sides of the casualty. The casualty's hands are then hooked around own neck and the casualty is pulled along, neck duly supported and the rescuer crawling on knees and hands.

When Two Rescuers are available : Many a time, a casualty cannot be carried by one person. In that case, more help must be sought. When there are two rescuers available, the casualty carriage is done through follwing methods:

● **A Four handed Grip :** For a sitting casualty, two rescuers hold each other's wrists in a manner that with right hand, the rescuers grip their own left wrist and with the free fingers of own left hand they catch each other's right hand. The casualty is made to sit and the casualty places own arms around the shoulders of the rescuers. There are many combinations of this grip. A three hand grip is also used with one of the rescuers using his/her free hand to support the casualty's back.

• A carriage with the help of a bed sheet: This may be the most commonly used methods for a casualty that can not sit on hand grips and has to be evacuated in laying position. See the picture.

When a casualty is to be carried on a stretcher, there are many methods available as shown in the pictures below. One must remember that the casualty should be safe and comfortable. Also, the stretcher should remain horizontal because of lack of space. In that case, the casualty has to be strapped to the stretcher and the head has to be kept up while lowering through narrow gaps. See the pictures carefully.

During Bomb Threats

9.15 Many a times, in a public transport or at a crowded place, unidentified and unclaimed packages are detected. These may be "innocent" packages or actually planted by anti-social elements. If such a package is found at a public place (and also a religious place), it results into a great panic and resultant explosion may lead to considerable damage. The following points should be noted :-

9.15.1 If found in a train :

• Alert the conductor and halt the train.
• Do not allow anyone to touch the package.
• Make sure that the passengers move away from the place of suspect package, to the next bogie in an orderly manner, without picking up their heavy luggage. Only small handbags should be picked up. Controlling this movement is very important. There should be no stampede. The people closest to the package should be moved first.
• The minimum distance of move should be at least 30 meters while the train is on the move.
• Alert the passengers of the next bogie too. Ideally there should be a gap of one bogie between the suspected package and the nearest passenger.

- After the train halts, make sure that the passengers are made to alight.
- Inform the nearest railway station.
- Inform the railway police travelling in the train and establish a cordon. Do not accept the requests of the passengers to allow them to pick up their heavy luggage.
- Get the bomb detection and disposal squad to defuse the bomb.

9.15.2 If found in a road transport.

- Halt the vehicle and off-load the passengers, without, picking up their heavy baggage. Control the crowd and do not allow any passenger to touch the package.
- Move the vehicle smoothly to a remote place, at least 50 meters away from the nearest building, vehicle, structure and in the open.
- Driver and cleaners / attendants must get down and take shelter away from the vehicle. Inform the police and amubulance services.

9.15.3 If found at a public place :

- Move the people out into the open and at least 50 meters away from the place / building in question.
- Cordon off the area and inform the police, fire brigade and the ambulances.
- Do not allow anyone to touch the package.
- Do not allow any one to re-enter the cordoned area.

9.16 There might be occasions when an unidentified caller informs of existence of a bomb in a premise. During such times, there is no point in making people evacuate the area without the knowledge of the exact location of the probable bomb. This is because, if the bomb actually exists, then one might unknowingly move from a safe area through a threatened area. For this, it is better to tell people to have a visual search of the area around them, without touching any unclaimed or a new package. Search techniques should be learnt by all. These techniques are simple and do not require much expertise.

9.17 After Bomb Explosions : Following actions should be taken :

- Inform the nearest police station/emergency police number and ambulance.
- Offer first aid on the spot - controlling bleeding and covering the burns.
- Evacuate to the nearest hospital that has adequate specialties available.
- Cordon off area and evacuate people beyond 100 mtrs.
- Do not crowd the roads and allow police, fire brigade and ambulance vehicles to reach the sport.
- Try to contain fire if any.
- Pull out trapped casualties from under the debris.
- Dividing the volunteers into groups - fire fighting group, casually evacuation group, rescue group and first aid group is essential, if every able-bodied citizen is trained in these basic skills, it would help mitigate the situation.

Search Techniques

9.18 When many searchers are available, searching of an area is quite easy. When searching, except for the searchers, other persons should be made to leave the place with their baggage. The most common methods are :-

9.18.1 Rabbit Hunt : This is the most manpower intensive method, but gives results is minimum time. The following diagram shows how an area is searched by the searchers.

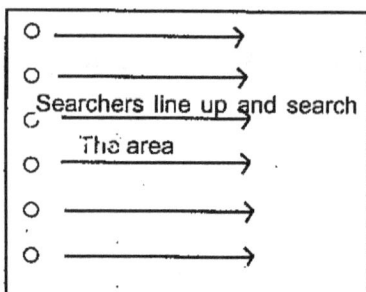

The searchers line up at hand-shake distance as indicated in the diagram. They note the presence of articles and look for things that are out of place and tampered. They do not open any bags or parcels, but the presence of articles is noted. Any presence of wire connected to parcels or ticking sound of a timer or any burning smell or any article that should not normally be present is seen.

9.18.2 Overlap Method : This method is used when lesser manpower is available. It is more time consuming. The method is shown in the following diagram :-

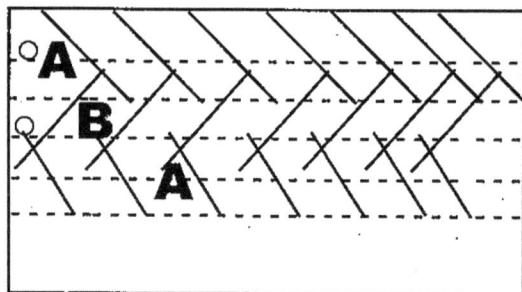

· Here, only two person are used and they undertake search of areas that are overlapping and are alternately covered, as shown in the figure. Thus, each area is searched by two persons in tandem.

9.19 There could be many methods of search and these will depend upon the area to be searched and the availability of manpower. Notwithstanding the methods used, each area of responsibility is vertically divided by each person and repeated rounds of search are taken. The vertical divisions are : Area from the floor to waist height, area from the waist height to the top of the head and the area above head level. Sticks, hammers and torches if available should be used. Search findings should be recorded.

9.20 There are methods of searching person and searching bags and parcels. However, these methods are not being given here because these require extensive training and if not correctly followed may lead to problems. Security persons should also take training in use of metal detectors and scanners. For any searcher, eye for details and the knack of finding the unusual need to be developed. The following principle should be kept in mind :-

No one is above board.

No amount of search is enough.

Remember how a place looked last time when you saw it.

Look for marks of disturbances.

Keep the size of a package and possible texture in view.

Look for smallest clues like hanging wires, uneven surfaces, change in textures and shapes.

Be alert to sound and smell.

Walk light footedly.

No vigorous shuffing of baggage should be undertaken.

Warning Note : Just by reading this book one cannot take adequate actions of offering first aid or perform casualty evacuation. It is strongly recommended that all persons should take formal training before attempting any of these methods. Following these methods without training and adequate practice may lead to unnecessary risks to the casualty. Thus, such a training should preferably be imparted in schools and colleges and repeated practices should be held.

LOWERING A STRETCHER WITH ROPES ONLY

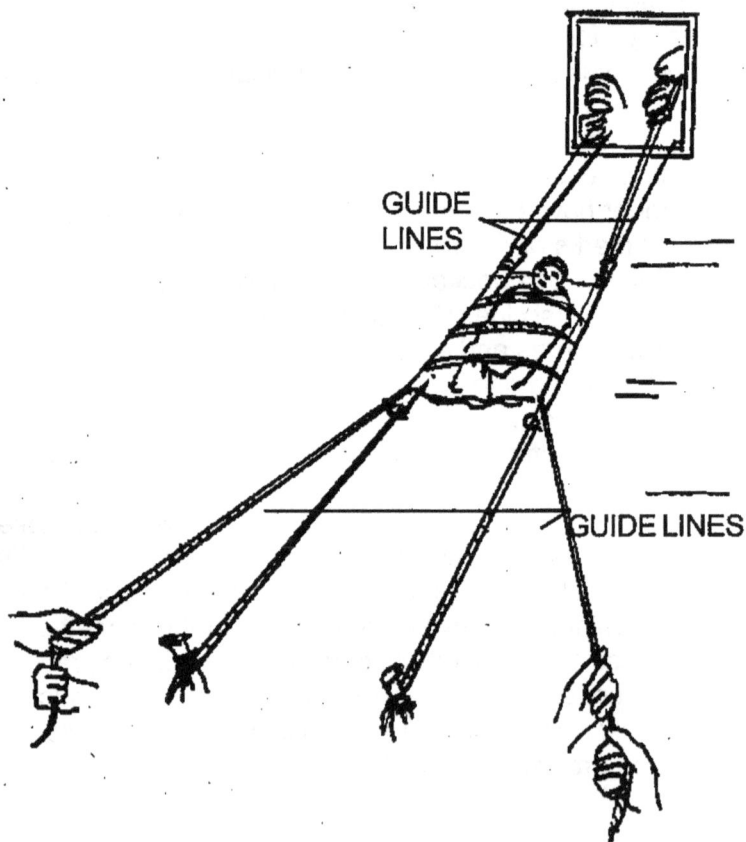

GUIDE
LINES

GUIDE LINES

LADDER HINGE METHOD

THE STRET CHER IS KEPT IN A
HORIZONTAL POSITION

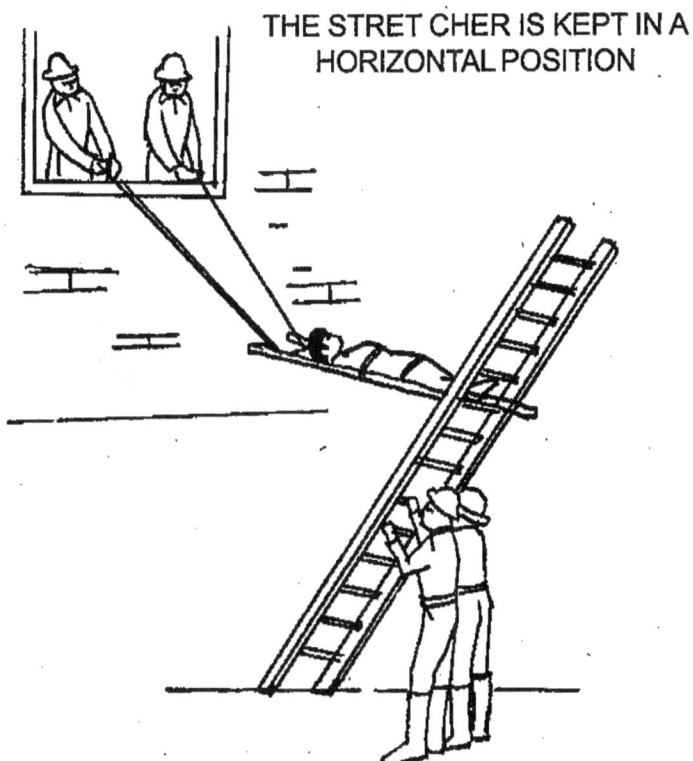

ANOTHER USE OF LADDER

THE STRET CHER IS KEPT IN A HORIZONTAL POSITION

SLIDING STRETCHER DOWN TWO LADDERS

FLYING FOX

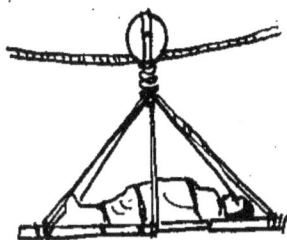

FIG. 1
RIGGING OF THE STRETCHER
FOR THE FLYING FOX

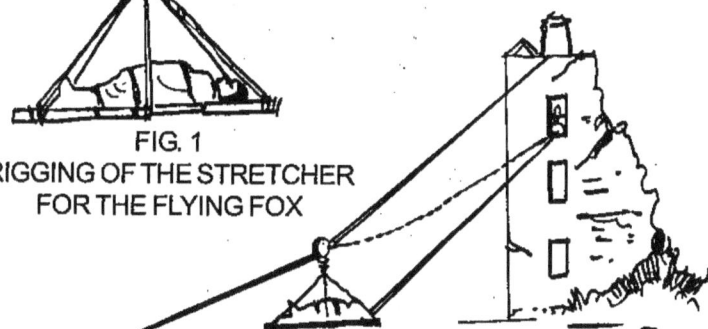

FIG. 2
THE FLYING FOX

RESCUE FROM
UPPER FLOOR

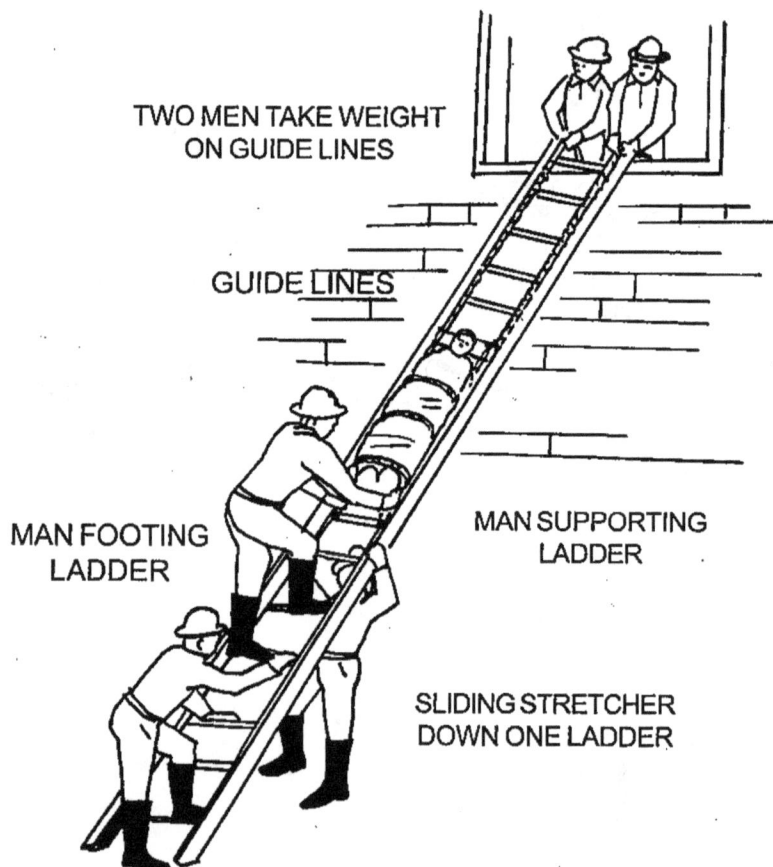

TWO MEN TAKE WEIGHT
ON GUIDE LINES

GUIDE LINES

MAN FOOTING
LADDER

MAN SUPPORTING
LADDER

SLIDING STRETCHER
DOWN ONE LADDER

TWO POINT METHOD

LOWE RING STRETCHER
FROM AN UPPER FLOOR

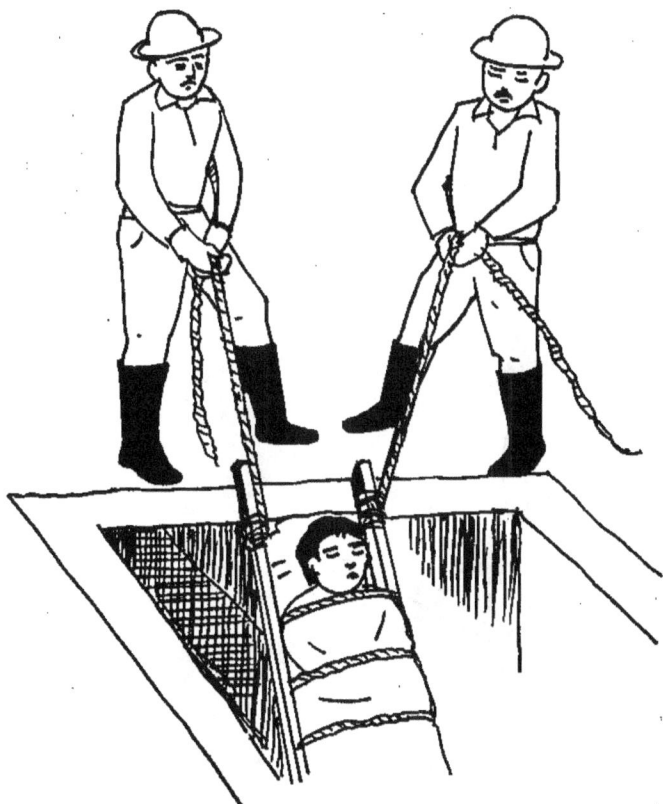

FOUR POINT METHOD

LOWERING STRETCHER
FROM AN UPPER FLOOR

EXAMPLE OF LEANING
LADDER METHOD

SUGGESTED PROJECT WORK

Project 1 : Visit a village that has experienced devastation due to floodsand conduct threat, vulnerability and rist analysis of the village. Students may have to spend 2 to 3 days and must study area maps, water channels, crops, building structures and interview village head and other citizen to learn about the previous disasters. This will help to analyse the future possible floods.

Project 2 : Visit a Taluka (Tehsil) headquarters and in conjunction with the Tehsildar and the BDO, make a draft plan for disaster mitigation to include the following :-

- Threat, vulnerability and risk assessment of different disasters.
- Resource availability for mitigation.
- Outline pian.
- Extra resource utilization.
- Information management.
- Communications.

This project may require 10 to 15 days of stay at the Taluka headquarters.

OTHER PROJECTS OF ACADEMIC NATURE

- Making disaster mitigation team at ward level.
- Creation of a disaster management cell at village level and writing down standard operating procedures.
- Creation of Taluka (Tehsil) level disaster warning systems out of local resources.

- Feasibility to organize a 'Warden Post' for a village with population of 10000 and with 25% of population rendered shelterless due to disaster.

- The duties that the NSS, Scouts and NCC cadets can undertake during, before and after the disasters.

- To work out village defence plan against terrorist activities.

- To create "general public intelligence and information system" to prevent disasters and take speedy mitigational actions.

❖ ❖

BIBLIOGRAPHY

1. Concepts and Practices in Disaster Management

                                    - Colonel (Retd.) P. P. Marathe

2. Readers' Digest

3. "Poonarjanma" (in Marathi)

                                    - Mr. Manohar C. Datar

www.ingramcontent.com/pod-product-compliance
Lightning Source LLC
Chambersburg PA
CBHW060604200326
41521CB00007B/660